AF190556

BrightRED Study Guide

Curriculum for Excellence

N5

DESIGN AND MANUFACTURE

Scott Atkins

BrightRED
PUBLISHING

First published in 2017 by:
Bright Red Publishing Ltd
1 Torphichen Street
Edinburgh
EH3 8HX

Copyright © Bright Red Publishing Ltd 2017. Reprinted 2019.

Cover image © Caleb Rutherford

All rights reserved. No part of this publication may be reproduced, stored in a retrieval system, or transmitted in any form or by any means, electronic, mechanical, photocopying, recording or otherwise, without prior permission in writing from the publisher.

The rights of Scott Atkins to be identified as the author of this work has been asserted by him in accordance with Sections 77 and 78 of the Copyright, Designs and Patents Act 1988.

A CIP record for this book is available from the British Library.

ISBN 978-1-906736-80-4

With thanks to:
Ivor Norman (editorial) and PDQ Digital Media Solutions (layout)
Cover design and series book design by Caleb Rutherford – e i d e t i c.

Acknowledgements
Every effort has been made to seek all copyright-holders. If any have been overlooked, then Bright Red Publishing will be delighted to make the necessary arrangements.

Permission has been sought from all relevant copyright holders and Bright Red Publishing are grateful for the use of the following:

Intel in Deutschland (CC BY-ND 2.0)[1] (p 10); Images licensed by Ingram Image on pages 11–12, 14–15, 20–21, 24–32, 35–39, 41–45, 53, 55, 60, 62–63, 67, 73–74, 77–81, 85 & 91–93; jean-louis Zimmermann (CC BY 2.0)[2] (p 14); osseous (CC BY 2.0)[2] (p 14); VFS Digital Design (CC BY 2.0)[2] (p 15); BorisPamikov/iStock.com (p 15); Juicy Salif designed by Philippe Starck. Used with permission of Alessi S.p.A, Crusinallo, Italy (p 15); Charles Chan (CC BY-ND 2.0)[1] (p 15); Liquid Light by Tanya Clarke copyright © 2017. Photography by Lisa Gizara (LiquidLightSite.Com) (p 15); Peter Shanks (CC BY 2.0)[2] (p 15); Low Voronoi Shelf by Marc Newson. Photography by Larry Lamay. Courtesy of Gagosian Gallery (p 15); Images by S. Atkins on pages 16, 17, 34, 48, 50–51, 54–62, 68–73 & 79; Jenny Cestnik (CC BY-ND 2.0)[1] (p 16); auroramixer (CC BY-SA 2.0)[3] (p 16); Reuben Flounders (CC BY-ND 2.0)[1] (p 19); Bonsoni.com (CC BY 2.0)[2] (p 22); Rino Peroni (CC BY-SA 2.0)[3] (p 23); 松林 L (CC BY 2.0)[2] (p 24); George Hodan (p 24); Cabanik (Public Domain) (p 26); SplitShire (Public Domain) (p 26); Maurizio Pesce (CC BY 2.0)[2] (p 28); Peter & Michelle S (CC BY 2.0)[2] (p 29); Omar Bárcena (CC BY 2.0)[2] (p 29); designmilk (CC BY-SA 2.0)[3] (p 31); DaveBleasdale (CC BY 2.0)2 (p 32); Charles & Hudson (CC BY-SA 2.0)[3] (pp 33 & 82); Quazie (CC BY 2.0)[2] and used with the permission of Inter IKEA Systems B.V. (p 36); Quazie (CC BY 2.0)[2] (p 36); Scott Moore (CC BY 2.0)[2] (p 41); James Wellington (CC BY 2.0)[2] (p 41); Anonimski (Public Domain) (p 41); Steve Bowbrick (CC BY 2.0)[2] (p 41); Karen and Brad Emerson (CC BY 2.0)[2] (p 41); Brodo (CC BY 3.0)[4] (p 41); Tenchi (Public Domain) (p 41); Alexi Kostibas (CC BY-SA 2.0)[3] (p 41); Das Ohr (CC BY-SA 3.0)[5] (p 41); Forever Wiser (CC BY 2.0)[2] (p 41); sarah (CC BY 2.0)[2] (p 41); Philipp Zinger (CC BY-SA 3.0)[5] (p 41); DC (CC BY-SA 4.0)[6] (p 41); Elke Wetzig (CC BY-SA 3.0)[5] (p 41); Bystander (CC BY-SA 3.0)[5] (p 41); Karpova Yana (CC BY-SA 3.0)[5] (p 41); Rotor DB (CC BY-SA 3.0)[5] (p 41); Tatesic (CC BY-SA 3.0)[5] (p 41); Andy Dingley (CC BY-SA 3.0)[5] (p 42); Douglas W. Jones (CC0 1.0)7 (p 42); Smoth 007 (CC BY-SA 2.0)[3] (p 42); Emilian Robert Vicol (CC BY 2.0)[2] (p 43); Audio-Technica (CC BY-ND 2.0)[1] (p 43); K P (CC BY 2.0)[2] (p 44); UnknownNet Photography (CC BY-SA 2.0)[3] (p 44); raeva/iStock.com (p 44); Windell Oskay (CC BY 2.0)[2] (p 44); Ozii45/iStock.com (p 44); Michael Pollak (CC BY 2.0)[2] (p 44); Pat Hayes (CC BY-SA 2.0)[3] (p 45); mamadela/iStock.com (p 45); Paul Downey (CC BY 2.0)[2] (p 45); Simon Eugster (CC BY-SA 3.0)[5] (p 46); Simon Eugster (CC BY-SA 3.0)[5] (p 46); Chris RubberDragon (CC BY-SA 2.0)[3] (p 46); Steve Payne (CC BY-ND 2.0)[1] (p 46); gotsumbeers (CC BY-ND 2.0)[1] (p 46); SpinningSpark (CC BY-SA 3.0)[5] (p 47); distelfliege (CC BY 2.0)[2] (p 47); Andrés Galeotti (CC BY 2.0)[2] (p 47); Double Happiness (CC BY-SA 3.0)[5] (p 47); Tenbergen (CC BY-SA 3.0)[5] (p 48); Johan (CC BY-SA 3.0)[5] (p 48); Andreas Mühlhausen (CC BY-SA 2.5)[8] (p 49); HarrisonB (Public Domain) (p 49); Plastic Coatings Ltd (p 49); OpenClipart-Vectors (Public Domain) (p 51); Torsten Henning (Public Domain) (p 51); Clker-Free-Vector-Images (Public Domain) (p 51); Duesentrieb (CC BY-SA 1.0)[9] (p 51); nickfrom (Public Domain) (p 51); Public Domain (p 51); Mass Communication Specialist 2nd Class James R. Evans/U.S. Navy (Public Domain) (p 51); Clker-Free-Vector-Images (Public Domain) (p 51); skeeze (Public Domain) (p 51); Martink (Public Domain) (p 52); Luke Milburn (CC BY 2.0)[2] (p 52); Paul N. Hasluck (Public Domain) (p 52); © CEphoto, Uwe Aranas (CC BY-SA 3.0)5 (p 52); Paul N. Hasluck (Public Domain) (p 52); Bigforrap (Public Domain) (p 52); SilentC (CC BY-SA 3.0)[5] (p 52); Scott Ehardt (Public Domain) (p 52); M. Minderhoud (CC BY-SA 3.0)[5] (p 52); Pitoutom (Public Domain) (p 52); Bob Key (Public Domain) (p 52); Za (CC BY-SA 3.0)5 (p 53); Za (CC BY-SA 3.0)[5] (p 53); Luke Milburn (CC BY 2.0)[2] (p 53); Setreset (CC BY-SA 3.0)[5] (p 53); Jameslwoodward (CC BY-SA 3.0)[5] (p 53); Aerolin55 (CC BY-SA 3.0)[5] (p 53); Jameslwoodward (CC BY-SA 3.0)[5] (p 53); Irwin Tools (p 53); Alexmadon (CC BY-SA 3.0)[5] (p 53); David J. Fred (CC BY-SA 2.5)[8] (p 55); William Warby (CC BY 2.0)[2] (p 59); Airman 1st Class Ryan Zeski/U.S. Air Force (Public Domain) (p 59); Phil Parker (CC BY 2.0)[2] (p 60); Fcb981 (CC BY-SA 3.0)[5] (p 62); 4tTrade/Travis Perkins plc for two images (p 62); Butch (CC BY-SA 2.0)[3] (p 62); Glenn McKechnie (CC BY-SA 2.0)[3] (p 64); Glenn McKechnie (CC BY-SA 2.0)[3] (p 64); MdeVicente (CC0 1.0)[7] (p 64); Simon Eugster (CC BY-SA 3.0)[5] (p 64); Jordanhill School D&T Dept (CC BY 2.0)[2] (p 64); Romary (CC BY-SA 3.0)[5] (p 64); Lucasbosch (CC BY-SA 3.0)[5] (p 64); MdeVicente (p 64); Audriusa (CC BY-SA 3.0)[5] (p 64); Evan-Amos (Public Domain) (p 64); M338 (Public Domain) (p 65); Zorro (Public Domain) (p 65); Simon A. Eugster (CC BY-SA 3.0)[5] (p 65); Satrughna (CC BY-SA 3.0)[5] (p 65); Jordanhill School D&T Dept for two images (CC BY 2.0)[2] (p 65); Flamefast (http://www.flamefast.co.uk/flamefast-ds430s-chip-forge-and-brazing-hearth.html) (p 67); Kentindustrialusa (CC BY-SA 4.0)[6] (p 68); Nottingham Hackspace (CC BY-SA 2.0)[3] (p 71); Glenn McKechnie (CC BY-SA 2.0)[3] (p 71); mark hobster/freeimages.com (p 72); FredFroese/iStock.com (p 72); Eyrian (CC BY-SA 3.0)[5] (p 74); HomeSpot HQ (CC BY 2.0)[2] (p 74); Kcida (Public Domain) (p74); JoJan (CC BY 3.0)4 (p 75); S.J. de Waard (CC BY-SA 2.5)[8] (p 75); Gareth Halfacree (CC BY-SA 2.0)[3] (p 75); S.J. de Waard (CC BY-SA 2.5) (p 75); CR Clarke & Co (UK) Limited for two images (p 76); 18004HENROB (Public Domain) (p 77); Creative Tools (CC BY 2.0)[2] (p 77); Dave Crosby (CC BY-SA 2.0)[3] (p 77); LaurensvanLieshout (Public Domain) (p 78); Robert Hewitt (CC BY-SA 3.0)[5] (p 82); gussencion (Public Domain) (p 82); Rohan von Indien (CC BY-SA 3.0)[5] (p 82); KUKKO (p 82); Glenn McKechnie (CC BY-SA 2.0)[3] (p 82); Simon Speed (Public Domain) (p 82); Dolev (CC BY-SA 3.0)[5] (p 82); Bushytails (CC BY-SA 3.0)[5] (p 82); Emrys2 (CC BY-SA 3.0)[5] (p 82); Emrys2 (CC BY-SA 3.0)[5] (p 82); Graibeard (CC BY-SA 2.0)[3] (p 82); Junkyardsparkle (Public Domain) (p 82); SilentC (CC BY-SA 3.0)[5] (p 83); U.S. Department of Defense (Public Domain) (p 83); Luigi Zanasi (CC BY-SA 2.0 CA)[10] (p 83) Photoblaz/iStock.com (p 84); Intel Free Press (CC BY-SA 2.0)[3] (p 85); supergenijalac/iStock.com (p 86); Sagadogo/iStock.com (p 86); FireAtDusk/iStock.com (p 86); Jordanhill School D&T Dept (CC BY 2.0)[2] (p 90); Filtre (Public Domain) (p 94).

[1] (CC BY-ND 2.0) https://creativecommons.org/licenses/by-nd/2.0/
[2] (CC BY 2.0) http://creativecommons.org/licenses/by/2.0/
[3] (CC BY-SA 2.0) https://creativecommons.org/licenses/by-sa/2.0/
[4] (CC BY 3.0) http://creativecommons.org/licenses/by/3.0/
[5] (CC BY-SA 3.0) https://creativecommons.org/licenses/by-sa/3.0/
[6] (CC BY-SA 4.0) https://creativecommons.org/licenses/by-sa/4.0/
[7] (CC0 1.0) https://creativecommons.org/publicdomain/zero/1.0/
[8] (CC BY-SA 2.5) https://creativecommons.org/licenses/by-sa/2.5/
[9] (CC BY-SA 1.0) https://creativecommons.org/licenses/by-sa/1.0/
[10] (CC BY-SA 2.0 CA) https://creativecommons.org/licenses/by-sa/2.0/ca/

Printed and bound in the UK by Ashford Colour Press Ltd.

CONTENTS

INTRODUCING NATIONAL 5 DESIGN AND MANUFACTURE

HOW WILL THIS BOOK HELP YOU?

This book has been carefully researched and designed to help you improve your chances of success within the National 5 Design and Manufacture course, and to act as a supplement to your learning in school. The key priorities when writing the book were to make sure that the material fully supported your learning, while helping you to secure success as you work through the course.

All of the content within the book has been matched to the key knowledge and understanding you are required to learn for the National 5 Design and Manufacture course. Throughout this guide, advice and useful tips on answering exam questions and completing coursework are given, along with extension materials including self-study resources, online tests, exam-style questions and video clips to help you develop and practise your skill and understanding across the course.

WHAT DO YOU NEED TO LEARN?

Each chapter within this book covers the content you need to know and to understand for the course. If you go through the Contents page, you will see that there are chapters covering each unit of the course as well as all of the key content within them. In addition to this, a Glossary of key terminology is provided at the end to help you easily find the information you are looking for. The content you must learn is broken down as follows:

Design

Within this subject area, you will cover the product-design process from the initial design brief to the final design of a feasible proposal (solution). Working through this unit, you are required to develop skills in, and understanding of, these aspects:

- the design process and its importance within society
- the role of design factors and how they affect design
- the need for research and the techniques required to carry out necessary research
- the need for specifications and how to generate them
- how to communicate initial ideas and develop them towards final proposals using idea generation, graphics and modelling techniques
- the design/make/test process
- the importance of evaluating and resolving design proposals on an ongoing basis, as well as the skills required to do this.

Materials and manufacture

Within this subject area, you will cover the product-design process from the final solution through to the manufacture of a prototype or product. This will help you to understand how to 'close the design loop' by taking a proposal forward for manufacture. Working through this unit, you are required to develop skill in, and understanding of, these aspects:

- the materials available for manufacture
- the equipment and processes used to manufacture using these materials
- the practical skills required in the accurate manufacture of items

contd

- how to plan effectively for manufacture by evaluating, refining and resolving plans
- the role of commercial manufacture in design
- the impact of manufacturing technologies on the environment and our society.

Further course and unit information, as well as a breakdown of the 'Course Specification' (this is the SQA's checklist of everything you must learn), can be found at the following link.

http://www.sqa.org.uk/sqa/files_ccc/DesignAndManufactureCourseSpecN5.pdf

 ONLINE

For all SQA documentation on the course, visit www.brightredbooks.net

ARE YOU READY FOR SUCCESS?

That's what this guide is all about. Don't feel daunted by the prospect of learning all of these new and unfamiliar terms and skills that the National 5 Design and Manufacture course will ask you to. Instead, use this guide to help you develop your understanding and to resolve any questions or uncertainties you may have. And, to make the process of preparation easier for you, every page has the following key features that you should look out for:

- **Don't Forget** text boxes that provide you with bite-size reminders of essential facts and tips.
- **Things to Do and Think About** text boxes require you to carry out some self-study to further develop your learning.
- **Online** links will provide you with further research and study materials.
- **Video Links** provide online links to videos to support your understanding.
- **Online Tests** provide you with access to tests that allow you to challenge yourself and assess how well you have learned topics.

Onwards and upwards

So, that's it. You are about to undertake new learning in a very exciting course that will provide you with a range of challenging and unique learning experiences. If you apply yourself and work hard, you will thoroughly enjoy the work being undertaken and will guarantee your success in achieving the National 5 award.

Good luck!

DESIGN AND THE DESIGN PROCESS

THE DESIGN PROCESS

WHAT IS DESIGN?

Everything not created by nature is the product of human design and engineering. We often take many of the everyday products we use for granted without ever really thinking about how and why these products are designed the way they are. So, what is design? Design can be described as the process of solving everyday human problems in our environment to help improve our lives and provide us with products that satisfy our needs and wants. All products large and small, from buildings to clothing, are designed. Design affects society and can influence our personality and lifestyle choices. Imagine your life without transport, smartphones and the internet. Many factors influence designs, such as fashion and technology – and, through constant innovation, designers can provide us with good design. However, designers can also create bad designs – and this is why it is important that a design process is followed to ensure the best possible outcome when solving any given problem.

THE DESIGN PROCESS

Stages of the design process

There is no 'one size fits all' model for design, as different design teams will approach design differently depending on the product being designed. However, we should realise that every design team will go through a series of stages between establishing an initial problem and realising and producing a final solution. Good design requires a design team to work between these stages, continuously evaluating each stage of the process and making any necessary changes as required. This should all be done in consultation with the different members of the design team as well as the client and consumers. The stages of a common design process, and a basic description of each, are given below. Some of these stages are explained in more detail later in this chapter.

Identification of a problem

At this stage, a problem or design opportunity is identified. This may be done by conducting a situation analysis, evaluating consumer needs and wants, or evaluating products.

Design brief

A design brief is the starting point of the design process. It outlines the problem, requirements of the solution and intended market.

Research

This stage requires investigation of the target market and design factors. This involves employing research methods such as surveys and product evaluations that will help to:

- determine what the consumer wants

- evaluate factors such as function, aesthetics and ergonomics to help you design a successful end product

- evaluate existing products and potential market competitors.

This research and investigation is very important, as, without it, the designer would be less likely to design a successful end product that would appeal to consumers and be able to compete with other products.

contd

Specification

A list of all required criteria for the final solution is generated.

Idea generation

Using idea-generation techniques as well as modelling techniques, the designer can begin to produce a range of ideas based on the requirements of the specification.

Idea development

After the initial ideas have been developed, the designer and design team will evaluate them to determine which ideas have the most potential. The ideas with the most potential will be developed, and any necessary changes will be made. Finally, the idea will be refined towards a workable solution.

Solution

The final concept is arrived at, and the proposal is presented to the client.

Planning for manufacture

Working drawings and manufacturing plans are drawn up that detail how the solution will be constructed. This will include details of materials and manufacturing processes.

Prototype development

A prototype is constructed. The prototype will be a fully working version of the final product.

Evaluation and testing

The client, consumers, design team and safety technicians can begin testing the product to see if it is fit for purpose. Any problems that arise will have to be solved. The design team will go back to previous stages of the design process to evaluate how best to resolve these problems. After this, they will apply these changes and test the product again until it is right.

Final manufacture and production

Once the product is finished and everyone involved is completely satisfied, the product can go into final production, ready to be sold to the consumer.

 ONLINE

Learn more about the design process by following the link at www.brightredbooks.net

 **DON'T FORGET**

Learn what happens at each stage of the design process, and relate this to the work you are producing in class. This will help you to answer questions about the design process in the exam.

 ONLINE TEST

Test your knowledge of the design team at www.brightredbooks.net

 ## THINGS TO DO AND THINK ABOUT

Watch the video at www.brightredbooks.net to give you an insight into how the design process works in action and what is being carried out at each stage. The stages in the video do not share the same names as those listed above. As you watch the video, consider where and when the stages listed above are happening.

THE DESIGN TEAM

INTRODUCTION

People often think of design as being the product of one person. However, working alongside the designer, there are a number of people who make up a design team. Each person has a specific job to do and is a specialist in their area. Although the designer will normally provide the vision for the solution, the rest of the design team must advise them on matters such as costing, manufacture and market to ensure the product is feasible and will appeal to consumers. Good design teams communicate effectively and will capitalise on the skills of each member to ensure the best design is achieved. The members of the design team listed below are instrumental in ensuring successful product development.

DON'T FORGET

You must learn the role of each member of the design team and understand how they relate to one another.

WHO MAKES UP THE DESIGN TEAM?

This diagram outlines the people who make up the design team.

Marketing team
Ergonomist
Manufacturer
Consumer
Designer
Market researcher
Retailer
Accountant
Economist
Engineer

Designer

The designer's job is to create and design solutions that are based on the needs and wants of clients and consumers. The designer will create ideas as well as detailed drawings and models that will help communicate their ideas to the rest of the design team.

Market researcher

Market researchers communicate with consumers, carrying out research to establish what it is they need or want. In doing this, market researchers can also help to identify gaps in the market. However, depending on the intended target market, it is important that the product is right for the consumer. Therefore market researchers have to research many of the following areas:

- current market trends, technologies
- consumer preferences, lifestyle, fashions
- demographics: age, gender, income, ethnicity
- geography: countries, cities, cultures.

By researching these areas, the market researcher can help ensure the designer gets the product right for its target market.

Accountant

Design projects come with budgets, as money is often required for resources that enable the different stages of the project to be carried out. The accountant will oversee project finances and provide advice on spending to ensure that the project stays within its budget.

Engineers

Engineers will advise the designer on the technical aspects of constructing and producing a product. They will have specialist knowledge of materials and construction. The engineer will also produce working drawings that will be used by the manufacturer to produce and assemble the product. Engineers are vital in helping to ensure the final product is safe and fit for purpose.

Manufacturer

Unlike the engineer, the manufacturer will provide the designer with advice on the manufacturing equipment and processes required to build the product. These processes will be chosen based on costing and volume of production.

VIDEO LINK

Watch the video clip at www. brightredbooks.net. This video, made by IKEA, will give you a good insight into how the design process works and how the key members of the design team work within it.

ONLINE TEST

Test your knowledge of the design team at www.brightredbooks.net

contd

Marketing team

The marketing team is responsible for promoting and advertising the product to ensure it grabs consumer attention. Marketing teams can do this by: creating adverts for media such as television or radio; producing billboard and magazine adverts; designing packaging; offering sale promotions; gaining celebrity endorsements; and designing store displays.

Ergonomist

Ergonomists specialise in providing information on human body sizes, physicality and psychology. This information is vital in helping designers to ensure their product is easy to use and safe for its intended market. See the Ergonomics pages (34–35) for a full description of ergonomics and its purpose in design.

Economist

Economists study the economy, and deal with market economics. This can involve looking at the current costing of competitor products and the cost of materials to ensure the product being designed is not overpriced for its intended market. Furthermore, economists will look at current issues such as the cost of living and income to help ensure that the product is affordable for consumers based on current wages and living standards.

Retailer

Retailers display and sell the product. They can provide useful information on current buying trends and consumer purchasing decisions. This information is valuable in helping to ensure designers get the product right. Furthermore, retailers can work effectively with the marketing team on how best to present and display the product to capture the attention of the consumer.

Consumers

The consumer is the end user of the product. Although they are not a direct member of the design team, they are one of the most important groups to consult. By communicating with the consumer, the design team can ensure their products fully satisfy the needs and wants of the consumer. If they do this successfully, then the consumer will purchase the product.

How do members of the design team link together?

This diagram shows how each member of the design team links with others during the product design process.

Designer:	Has to work with everyone to ensure the best possible outcome is reached.
Market researcher:	Designer, Marketing team, Consumer, Retailer.
Accountant:	Designer, Manufacturer, Economist, Retailer.
Engineers:	Manufacturer, Designer, Economist.
Manufacturer:	Designer, Accountant, Engineer, Economist.
Ergonomist:	Consumers, Designer, Market researcher.
Economist:	Manufacturer, Designer, Accountant.
Marketing team:	Retailer, Designer, Consumers, Market researchers.
Retailer:	Marketing team, Designer, Consumers, Market researchers.
Consumer:	Retailer, Marketing team, Market researcher.

THINGS TO DO AND THINK ABOUT

Copy the diagram which shows the links between members of the design team. Add a new column that outlines what may be discussed and what activities may be carried out when these members of the design team communicate with one another.

PROBLEM IDENTIFICATION, DESIGN BRIEFS, RESEARCH AND SPECIFICATIONS 1

PROBLEM IDENTIFICATION

The first stage of design is to identify a reason to design something. Designing and developing products costs money, therefore research has to be conducted to find potential markets in which to develop new products. Markets for new products can be identified in the following ways:

Developing new products

When new technologies or manufacturing techniques become available, designers can use these to develop new products that haven't been seen before. This can capture the market and push design onto the consumer. *(See 'Technology push', pp. 28–29.)*

Conducting market research

Finding out what consumers need or want is crucial in discovering opportunities for product development. By carrying out research, we can find gaps in the market, assess consumer demand and develop products that meet people's needs/wants. *(See 'Market', pp. 26–29.)*

Product evaluations

This involves evaluating existing products to find ways of improving them or establishing new and better products. Key design factors such as function and aesthetics may be evaluated to find ways of improving functionality and styling. This can help in developing products that effectively compete at market and appeal to the consumer.

Situation analysis

A situation analysis is a way of analysing the internal and external factors surrounding a market or a product to help identify new opportunities. There are several ways to carry out a situation analysis – and one of the most common is **SWOT analysis**. SWOT analysis evaluates internal **S**trengths and **W**eaknesses and external **O**pportunities and **T**hreats. For example, if a company wanted to build upon their existing range of successful laptops, they may consider the following:

- **S**trengths: Established product that is selling well and where customer feedback is positive.
- **W**eaknesses: Product has been on the market for too long without any new developments.

- **O**pportunities: New technologies are available that could improve the design and renew interest, i.e. touch/detachable screens.
- **T**hreats: Other companies are also developing these products, or the market may not be ready for them.

The situation analysis in this instance identifies problems with the current laptop range and identifies areas where a new opportunity may be held. Further research may lead to the incorporation of technologies such as detachable touch screens.

DESIGN BRIEFS

Once a problem has been identified, a design brief is produced. A design brief is often the starting point of the design process. A design brief will outline the problem, provide the designer with an indication of what must be done, and set down any restrictions that must be adhered to in the design. Design briefs often take two forms:

- **Open briefs** are very open to interpretation and will have few restrictions placed upon the designer. Open briefs allow the designer to be more expressive and have more say over the final design.
- **Closed briefs** are restrictive. They will normally tell the designer what must be adhered to and will limit the amount of say a designer will have in the final design.

 contd

Design-brief analysis

After the design brief is produced, the design team will analyse all of it. This normally takes the form of a mind map and seeks to identify everything to be done, including areas for further research. Here is an example design brief:

Problem: We are a popular restaurant chain looking to expand into the student market, offering fast-food restaurants with affordable produce and a relaxed setting.

Brief: We would like you to design seating and tables that would be suitable and appealing to the student market.

The analysis of this brief may include the following:

VIDEO LINK

Watch the video clip at www.brightredbooks.net for a good insight into the need for design briefs.

Ergonomics
- How can we make the seats comfortable to sit on?
- What size should the seating/tables be to accommodate a range of people?
- How much space do people need to eat/work comfortably?
- How will we make it easy to get in and out of the seating?
- Can the furniture be moveable? If so, how heavy, stable and easily moveable will it have to be?

Function
- Could the seating and tables offer more functions, such as for studying?
- How can the seating/tables be laid out to make socialising easier?
- Should the seating/tables be adjustable or moveable to maximise comfort?
- Should a secondary function of the seating/tables be aesthetics to fit in with the restaurant's styling?

Restaurant furniture

Materials
- What materials can be used to ensure the seating/tables are strong, hard-wearing and easy to clean?
- How can we use materials to enhance the aesthetics?
- What materials can be made stain-resistant?
- What materials can be used sustainably?
- How can we reduce the cost?

Aesthetics
- What styles/fashions do students like?
- What themes appeal to students?
- What styles do other fast-food chains use?
- What styles does our fast-food chain use?
- How do we make it appeal to students without making the appearance look cheap or too upmarket?

This basic example shows some of the questions or aspects the design team will have to consider. Once these are identified, research can begin to find answers to these aspects so that an effective solution can be developed.

 ONLINE TEST

Test your knowledge of the design team at www.brightredbooks.net

 ONLINE

Learn more about design briefs by following the link at www.brightredbooks.net

RESEARCH

After a brief has been written up, the design team will begin conducting research. Remember that research can also be carried out before the brief to help to establish new markets and products for development. Market research helps to:

- establish what it is consumers need/want
- find potential gaps in the market
- assess current products on the market to see what they offer
- establish areas for improving these products
- gain data on how to best design your product.

During the design process, designers will also look for research to be carried out during the ongoing development of ideas. This ensures that any products being developed will meet the requirements of the **client**, **consumer** and **design specification**. There are many evaluation techniques that can be used to conduct research. (see next page).

 DON'T FORGET

You must understand and be able to explain the purpose of design briefs and how to analyse them.

 ## THINGS TO DO AND THINK ABOUT

1. Describe three ways in which new product developments can be identified.
2. Explain the purpose of the design brief and brief analysis.

PROBLEM IDENTIFICATION, DESIGN BRIEFS, RESEARCH AND SPECIFICATIONS 2

DON'T FORGET

You must understand and be able to explain how to use questionnaires and user trips, as well as the advantages they offer the researcher.

RESEARCH TECHNIQUES

The following two research techniques are often used to conduct research as part of a design brief analysis.

Questionnaires/surveys

Questionnaires/surveys are a good way of collecting a large number of responses quickly. They are useful when evaluating factors such as **aesthetics**, where pictures can be shown and questions asked to determine the consumer's preferences with regard to appearance and styling. We can also evaluate factors like **function**; however, doing this in depth would be difficult via a survey, as the user cannot physically use the product and can only guess how it works from a picture and questions.

Once the design brief has been analysed and key areas of research considered, the marketing team will write up questions to distribute to consumers. These questions will seek to answer the issues outlined in the brief analysis. Survey questions must be well considered to ensure the responses collected are useful and give us the information we require. For example:

- *Do you like the kettle shown?* **Yes** or **No.**

This only tells us whether or not a user likes the kettle and nothing specific about its aesthetics therefore our results are limited.

- To improve responses and get more information, we may include pictures of several kettles and ask the consumer: *On a scale of 1–5, with 5 being excellent, how would you rate the colour scheme of each of the following kettles?*

- We could go further by asking the user to provide reasons to the previous question to gain more insight into consumer opinion.

- This response will ensure the data generated from the survey provides us with useful information we can take forward into the design stage.

User trips

A user trip involves the designer physically using and testing a product. This design analysis can relate to points highlighted from a brief analysis and will help the designer to determine potential issues/areas for improvement for the product in question. User trips can also be used to identify market opportunities for new product developments.

To carry out a user trip, the designer will outline specific activities that are to be carried out in relation to the users in question. The designer may then consider questions such as:

- Who will be using the product and what issues may occur for different types of users?

- What does the user want the product to do?

- How will the product be used?

- How will the product perform when carrying out specific functions before, during and after use?

- In what sort of environment will the product be used?

- What are the priorities when using the product? For example, which functions will be used most often?

contd

The designer can record this information in several ways such as note taking, audio recordings and/or photographic and video evidence. The data can then be used to drive the design of the product during the design phase.

SPECIFICATIONS

After research has been carried out and conclusions have been drawn from this research, a design specification is written up. A specification is a list of the things the product must do. This will then be discussed with the client to ensure they are happy with what the specification sets out to achieve. Once agreed, the specification can then be used to begin generating ideas that solve the brief. The specification may provide specific information on a variety of design factors based on the conclusions drawn from the design brief analysis, questionnaires and user trips. This includes function, styling (aesthetics) and materials to be used. For example, research of function and aesthetics for the design of a new child's garden play zone may result in the following specification:

1. Function

1.1 must fit within the average back garden.

1.2 must include slides and climbing areas.

1.3 must incorporate a learning feature.

2. Aesthetics

2.1 must be colourful and appealing to children.

2.2 must use simple shapes that are easily understood and appealing to children.

3. Materials

3.1 must be durable, hardwearing and withstand varying weather types.

3.2 must be suitable for children to ensure the product is safe.

Obviously, the above specification is a basic example and specifications in the real world will be much more detailed. They will likely consider many more design factors and include information that details key decision made from research that has been carried out.

DON'T FORGET

You must be able to describe the role of a specification and how they are generated using data from research. You will also be required to generate specifications during your coursework. Use this experience to help you fully understand this topic.

DON'T FORGET

You can be asked to come up with specification points in exam questions so be prepared.

ONLINE TEST

Test your knowledge of problem identification, design briefs, research and specifications at www.brightredbooks.net

THINGS TO DO AND THINK ABOUT

Think about three products you know of or use a search engine to find products. For each product:

1. Think about what conclusions may be drawn from the research techniques mentioned previously.

2. Now write down a specification that outlines what you think would have been considered before designing the product.

IDEA GENERATION

After a specification has been written up, the next stage is to begin generating ideas. It is important that ideas created meet the specification. However, coming up with ideas is not always easy. To help with this, there are a variety of ways in which the designer can generate ideas.

DON'T FORGET

You must know how to describe and use brainstorming and morphological analysis for your exam.

THOUGHT SHOWERS (BRAINSTORMING)

Thought showers (brainstorming) is a way to generate ideas where an individual or members of a group quickly and spontaneously come up with ideas for solutions good and bad. In a group, a team leader often leads the discussion and records all of the thoughts and ideas given in relation to potential solutions for a problem. A spider diagram or mind map is often used to lay out and visualise a brainstorming activity with annotations and sketches.

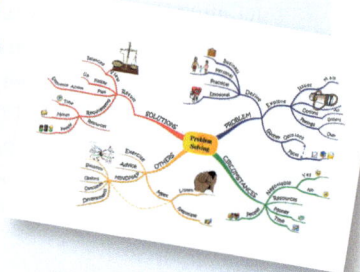

Once complete, all ideas are evaluated and the best ones taken forward. These are then turned into potential ideas and developed via sketching and modelling techniques.

MORPHOLOGICAL ANALYSIS

This technique involves breaking down the problem into different categories based on aspects such as function that will affect the product's design. These categories are included as headings in a table, and then sub-categories are included below them. The sub-categories are then randomly mixed to generate potential ideas. Here is a basic example of a morphological analysis for an office chair:

Stability	Style	Materials	Features
1 leg	Modern	Hardwood	Curved for body
2 legs	Traditional	Coloured plastic	Arm rests
4 legs	Organic	Metal	Cushioning
Solid base	Standard design	Softwood	Wheels

By mixing the sub-categories as shown in the table, the chair shown could have been created. One of the benefits of morphological analysis is the number of variations it can create to help you to generate ideas. In the example above, 256 variations can be found, providing you with numerous ways to design your ideas.

TECHNOLOGY TRANSFER

Technology transfer, as its name suggests, is the transfer of technology from one product to another. New technologies and materials are being researched constantly, and new discoveries generate ideas for new products. For example, NASA's research into protective padding to improve crash protection for astronauts and for aeroplane passengers resulted in the development of 'memory foam'. Designers used this technology to generate the idea of memory-foam beds to improve sleep and to ease pain for people with muscular/back problems.

ONLINE

Learn more about other NASA developments that have created technology-transfer ideas by following the link at www.brightredbooks.net

MOOD/LIFESTYLE BOARDS

Mood boards are used to collate a series of images that help to visualise the theme surrounding a product. This could involve looking at colours, shapes, location/ environment, products and patterns that relate to the product to be designed, in order to provide inspiration/direction for a product's styling.

contd

Example of a mood board

Example of a lifestyle board

Lifestyle boards are used to collate images that help to visualise the target market. This could involve looking at consumers' interests, housing, fashion, food choices, likes, dislikes and music tastes. Using these images, the designer can then begin sketching ideas, incorporating the ideas presented onto the mood/lifestyle boards.

ANALOGY

Analogy is a way of generating ideas by considering how other products, whether manufactured or natural, can inform your design ideas. When Philippe Starck designed the highly regarded 'Juicy Salif' lemon-squeezer, he stated that the idea came from squid, which he was eating at the time.

Many other products have also been designed in this way.

Squid into a juicer, Juicy Salif, designed by Philippe Starck

Running tap into a table light, Liquid Light, designed by Tanya Clarke (LiquidLightSite.com)

Honeycomb into a bookshelf, Voronoi Shelf, designed by Marc Newson

LATERAL THINKING

Lateral thinking is best described as 'thinking outside of the box'. This involves looking at the problem in a completely different way and trying to design ideas that break the norm. For a time, communication between people in different countries was done via mobile phones, which cost money, and via e-mails/messaging, which limit live social interaction. Skype video calling, which combines elements of each as well as video technology, was the product of lateral thinking, and changed how people interact across the world.

THINGS TO DO AND THINK ABOUT

In this course, you will have to demonstrate the ability to generate ideas. Try carrying out some of the techniques above for a product of your choice. Then, using the sketching techniques outlined on the next spread, try to create a range of diverse ideas incorporating the findings from your idea-generation techniques.

ONLINE

Learn more about idea generation by following the link at www.brightredbooks.net

ONLINE TEST

Test your knowledge of idea generation at www.brightredbooks.net

GRAPHIC AND PRESENTATION TECHNIQUES

GRAPHICS TECHNIQUES

ONLINE

Follow the link at www.brightredbooks.net to a website which discusses the importance of sketching in design.

ONLINE TEST

Test your knowledge of this topic at www.brightredbooks.net

To help visualise ideas fully, designers will normally use a range of graphics techniques to communicate their ideas. There are various methods the designer can utilise to do this:

2D sketches

Two-dimensional sketches can be used to show the whole product or individual component parts of the product. These can allow the designer to show different views of the front, side, top and rear in 2D. This is also referred to as **orthographic** sketches. Orthographic drawing is covered later on pp. 50–51.

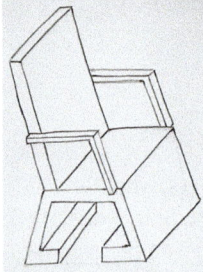 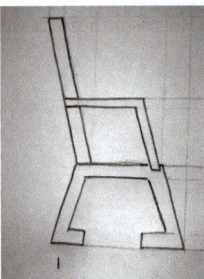 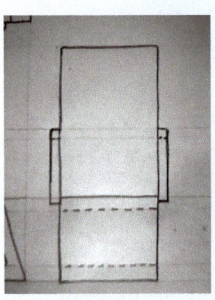 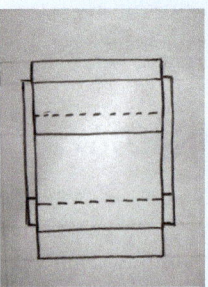

Oblique view   2D side view   2D front view   2D top view

2D sketches of a chair design

3D pictorial

Three-dimensional sketches can help us to better visualise the form of our designs, making them easier to understand and demonstrate refinements as we develop our designs. Different methods of 3D pictorial sketches can include:

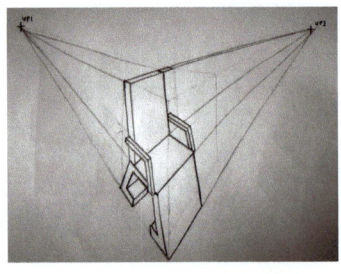

1-point perspective example

- 1- and 2-point perspective (realistic depiction of distance and depth). Lines are drawn back to 1 or 2 vanishing points that lie on a horizon line. This technique creates a realistic image of what our eyes see, where items seem smaller the further away they are from us. 1-point can be easier to sketch as we only have to start with a 2D drawing before taking lines back. However, for certain objects, 2-point is better, as it helps us to visualise and show all sides of the object more clearly.

- Isometric (shows all 3 sides of an object clearly). Lines are drawn back at 30°.

- Oblique (shows the front clearly, with added depth). Lines are drawn back at 45° making this technique useful for anyone struggling to sketch in 3D.

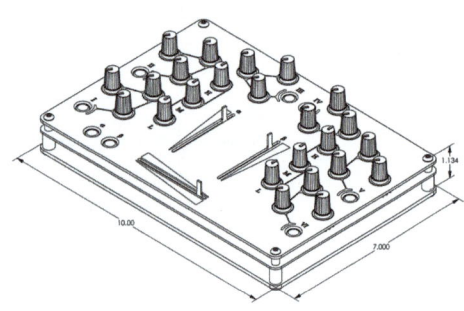

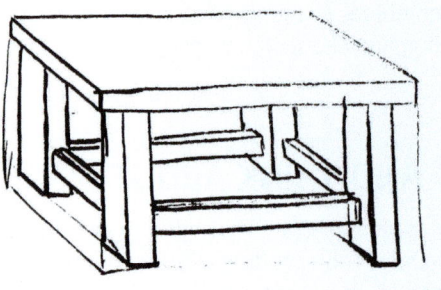

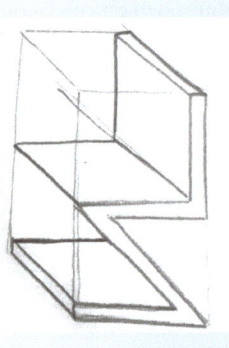

Isometric example   Oblique examples

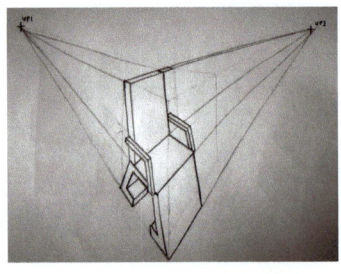
2-point perspective example

contd

To further detail the workings of the idea, the designer may include exploded views or sectional views:

Exploded views

Individual parts are separated and drawn in line with one another to show how the product is to be assembled. This can be done in 2D or 3D and is extremely useful when trying to explain how a design will work in terms of manufacture and assembly. Exploded drawings can be done using technique in 2D or 3D.

Partial Enlargements

These can be used to zoom in on complex features, providing clearer detail of how components work or are assembled.

Sectional views

Sectional views display a cut through a product to make the internal workings of a product clearer. This can be useful when showing how parts are assembled or to make complex details clearer. This can be done in 2D or 3D.

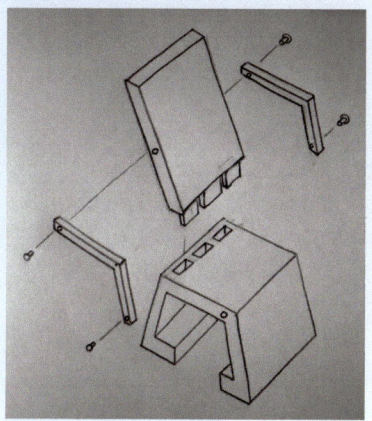

Exploded view of a chair design

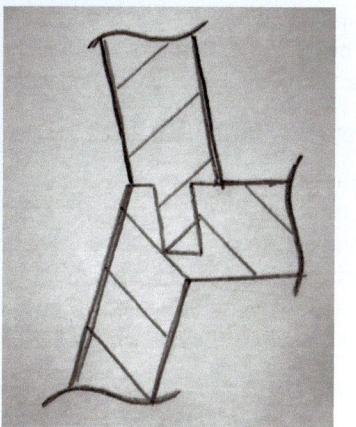

2D sectional end elevation of a chair design

ILLUSTRATION

Once the designer has sketched their ideas, they can then illustrate (**render**) them to make the sketch look more realistic, adding material textures, tone, light, shadow and reflection while also considering potential colour schemes. Three of the most common rendering mediums are:

The choice of medium for rendering is often determined by the type of product. For example, a product that requires high impact would often be rendered using marker pens. Car designers often use this technique.

VIDEO LINK

The video at www. brightredbooks.net demonstrates how sketching and rendering can be used to visualise your ideas.

Pastel: for soft colour and tone

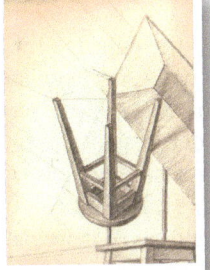

Pencil: coloured or monochrome, general use

Marker pens: for bold colour and tone.

Annotation

To help further communicate ideas, designers often annotate their ideas and design developments. Annotations normally take the form of labels or short notes that discuss design issues and explain the functionality of ideas.

DON'T FORGET

You must be able to explain the purpose of each type of graphic listed here and how these can be used to generate/develop ideas and communicate designs.

 ## THINGS TO DO AND THINK ABOUT

Go online and search for videos on the sketching and rendering techniques outlined above. You will find good examples and tutorials that will allow you to practise and develop your understanding of these techniques.

Buy yourself a sketchbook, and sketch regularly at home to help you develop your skill in sketching. Illustrate your sketches as well.

IDEA DEVELOPMENT AND MODELLING TECHNIQUES

IDEA DEVELOPMENT

Once ideas have been generated, the next stage is to develop the best ideas towards a final design proposal. Many of the graphics techniques outlined previously are used when developing ideas to effectively communicate what changes are being made and how this will improve the design. Annotations should also be included, as well as justifications that explain exactly why you have decided to develop an idea in a certain way. This may be a justification that explains how the development fully meets the specification or further solves/improves design issues such as **ergonomics** or **function**.

What ideas to develop?

Developing ideas requires you to evaluate and think about what improvements or modifications are required to improve your design ideas and turn them into potential solutions. A good way to do this is to review all of your ideas, including your annotation of design issues, and then use an evaluation technique to assess how well each idea solves each specification point from your specification. One technique is to use an evaluation matrix such as the one shown here.

An **evaluation matrix** allows you to quickly identify the strengths and weaknesses of each idea, where a tick or cross is used to indicate whether the idea satisfies each specification point. Evaluation matrices also display the overall total for how well an idea meets the specification, allowing you to see which ideas are the strongest to take forward. It is important to note, however, that even though an idea doesn't meet all of the specification points, you can still take it forward. This may be because you feel the idea has good potential and that, with further development, it could become a viable solution that fully meets the specification.

IDEAS	1	2	3	4	5	6
SPEC POINTS						
1.1	✓	✓	✓	✗	✗	✓
1.2	✗	✗	✗	✗	✓	✓
2.1	✓	✓	✓	✗	✓	✓
2.2	✓	✗	✗	✗	✗	✗
3.1	✓	✓	✓	✓	✗	✗
TOTAL (5)	4	3	3	1	2	3

Evaluation matrix

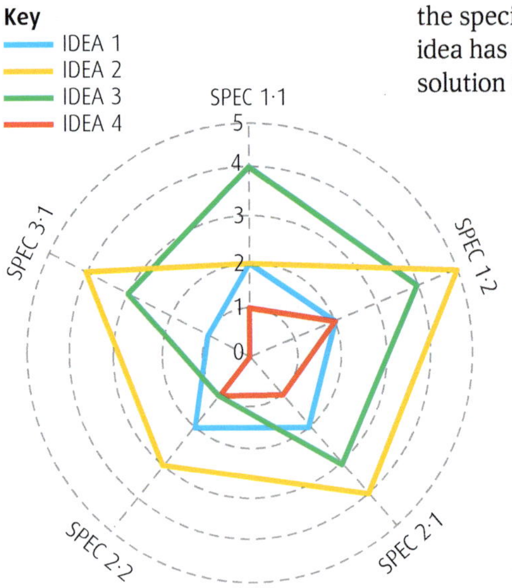

Another technique is **spider-web analysis** such as shown here.

Spider-web analysis allows you to further grade each idea against each specification point, as is shown here. Ideas are graded 0 to 5 from the centre to the outside of the web. Each grade is indicated as a circle. Specification points are then listed around the outside, and a line is used to connect them back to the centre, creating a web. A line can then be drawn to grade each idea against the specification. An idea's strengths and weaknesses are easily assessed by how far the line representing an idea stretches to the outside of the web. This can be a useful technique for evaluating each individual idea; and a spider-web analysis can be carried out beside each idea.

Key
- IDEA 1
- IDEA 2
- IDEA 3
- IDEA 4

Begin design development

You should think of development as an exploration of your chosen ideas. This can include:

- **Synthesis of ideas:** the combination of the strongest elements from different ideas to produce a new idea that fully solves the specification. However, great care has to be taken when doing this to avoid making random developments that are actually worse than the original idea. All synthesis should be based on solid reasoning that has come from your evaluation of ideas.

- **Idea exploration:** exploring an individual idea to improve its areas of weakness, looking at potential ways to solve design issues. Further research may have to be carried out to help you find solutions.

- **Refinement:** refining the idea, looking at enhancements you can make with regard to design issues to ensure the idea fully satisfies the brief and specification.

MODELLING TECHNIQUES

To help further communicate the development process, designers will often use models. This can help the designer to physically see and gain a better idea of the proportions of the design. It can also allow testing to be carried out, testing materials, function, ergonomics and safety. It is important to note that modelling can also be used during the initial ideas stage.

Model types

There are various types of models, and you must be aware of each and when they are used.

- **Sketch model:** a quick, easy to make and inexpensive technique where a 3D model based on your initial sketches is produced to help you visualise the shape and form of an object in physical 3D. Sketch models are rarely to scale and are often made from paper, card or thin foam boards. They allow you to quickly communicate and evaluate aesthetics making edits as required to improve designs.

- **Block model:** a model, as its name suggests, made from one block of material. Unlike the sketch modelling, a block model is more accurate and focuses directly on the physical appearance of the product. A block of material is shaped and prepared to represent the physical appearance of the product and will contain no internal components. Block models are very useful in evaluating and editing aesthetics and function to improve your designs.

- **Scale model:** a physical model made full-scale or to a portion of the actual size of the final design. These allow the designer to evaluate and determine final sizes and also see how the model works ergonomically to ensure the design will work for the intended end user. Scale models require a reasonably good level of skill to make, and the cost can vary depending on material choices.

- **Computer-generated model:** a 3D computer model made using CAD software. This can be used to accurately reflect aesthetic choices, to test safety issues through computer simulation and to determine sizes for the final design. A software specialist is required to design the computer model. The cost of making the model is low, but the computer hardware and software required are expensive. The downside to computer models is that they are not physical, which limits some of the testing that can be carried out.

- **Prototype:** a physical full-scale working version of the final design using exact or as close to the final material choices as possible. Prototypes function exactly as the final product would and allow full testing of the final design. These are very expensive to make and require the expertise of manufacturers and engineers to create a working final design.

Modelling materials

Models can be made from almost anything. The key to modelling is to try to create a model that accurately represents the design you are modelling. Obviously, the materials chosen should reflect the intended purpose of the model, as is outlined above. The following materials are commonly used in model-making due to their good workability:

- paper
- card
- corrugated card
- metal wire
- pipe-cleaners
- foam
- expanded foam
- plastic sheets
- MDF
- balsa wood
- modelling compound
- modelling clay
- smart materials
- construction kits.

Computer-generated model of a loudspeaker design

DON'T FORGET

You must know how sketching and illustration, as well as modelling techniques, can be used to inform the development of ideas.

DON'T FORGET

Learn each modelling technique and be able to explain how to use each technique and the benefits of doing so.

ONLINE

Learn more about modelling by following the link at www.brightredbooks.net

VIDEO LINK

Although this video focuses on prototyping, it will give you a good insight into why we need models and how we make them. Watch the video clip at www.brightredbooks.net

ONLINE TEST

Test your knowledge of development and modelling techniques at www.brightredbooks.net

THINGS TO DO AND THINK ABOUT

Using the internet and YouTube, research the modelling types and materials listed above to help you fully understand modelling. You should make your own notes as you do this.

DESIGN FACTORS

INTRODUCTION

CONSIDERING DESIGN FACTORS

Design factors are extremely important. They allow designers to consider the many aspects and issues that will affect the outcome of a solution for any given product. Effective consideration of these factors will ensure that the end product is a success. No single factor is more important than any other, as each and every product will require specific factors to be considered. For example, if we take a high-end, high-specification, one-off design for a sports car, **cost** will not be an important issue – whereas, if we were to consider the design of a more affordable mid-range sports car, cost will be a very important factor in ensuring that the consumer will want to pay for the end product.

There are several factors that must be considered during the design process. There are five key factors that you must learn for your National 5 exam, including the individual aspects of each. These are as follows:

- **function**
- **market**
- **ergonomics**.
- **performance**
- **aesthetics**

These are all explained in more detail later in this chapter, and you must be familiar with the key aspects of each. Aside from these five key factors, you must also be able to discuss safety, cost, sustainability and the environment in relation to design and manufacture.

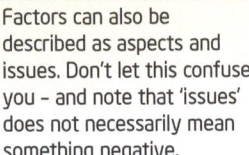

DON'T FORGET

Factors can also be described as aspects and issues. Don't let this confuse you – and note that 'issues' does not necessarily mean something negative.

SAFETY

All products should be safe to use for the consumer. Designers and manufacturers all work with given regulations to ensure that a product meets considerable safety standards before being placed on the market. In addition to this, products go through rigorous testing to ensure they meet these safety standards so that we ultimately limit the chance of injuring, straining or even causing the death of the consumer.

When undertaking coursework you may have to evaluate safety and, in the exam, you could be asked to discuss how safety has influenced the design of a product. An example to help you with this is as follows.

A toddler's bottle is shown here.

How has safety influenced the design of the toddler's bottle?

The materials should be durable, as toddlers may throw, drop or bash the bottle. The handles should be easy to hold and grip so that the toddler can use and hold the bottle easily and safely. The liquid should not spill out of the bottle if tipped, as this could create a slip hazard for parents and the toddler. If the toddler chews the drinking spout, no parts should break off, to avoid choke hazards.

DON'T FORGET

When answering questions on safety, do not use the answer 'no sharp corners'. Instead, try to think about the specific safety issues that could be a risk to the user with regard to the given product.

COST

Cost can be an important issue to consider in product design. When products are being developed, companies often have a budget to consider. To ensure they stay within this budget, they must consider the economy and what type of market a product is aimed at. Furthermore, other factors such as the cost of raw materials, fixture and fittings, transport logistics, packaging and standard components will affect cost.

contd

When undertaking coursework you may have to evaluate cost and, in the exam, you could be asked to discuss how cost has influenced the design of a product. An example to help you with this is given below.

> An economical design for a hand blender aimed at a mass market is shown here.
>
> How will cost have affected the design of the blender?

As the product is low-cost, the plastics used would have to be readily available and cheap. The product would also have to be mass manufactured to reduce the cost of manufacturing and the end price to the consumer. Standard components such as the blades and electrical parts would also have to be low-cost and bought in bulk to reduce the cost.

THE IMPACT OF DESIGN ON THE ENVIRONMENT AND SOCIETY

With new developments in technology and manufacturing, new products are constantly being produced. As consumers, we are continually buying and demanding these to improve our lives and satisfy our needs and wants. This is the **rise of consumerism**.

In addition to this, we are demanding newer products that represent value for money and that are lower in cost. We often buy these products without any consideration for the environment or our society. Demanding continual product development and cheaper products can result in many issues such as:

- products being made in sweatshops using lower-paid workers with less access to workers' rights and safe working conditions;
- products being made in countries with limited regulations over waste disposal from manufacturing;
- old products being replaced unnecessarily;
- newer versions of products being made because technology is constantly changing. Not every new version marks a significant development in the product, so do we need to buy it? It could be that our only reason for purchasing it is because of social and peer expectations.

As consumers, we must consider these issues when making purchasing decisions. Furthermore, we must also consider the sustainability issues with products. Products that are sustainable are produced on an understanding that they will not damage the environment unnecessarily. This can include:

- using materials that are sustainable, recycled or upcycled (see the Materials chapter for more on this);
- using transport logistics that reduce carbon emissions (for example, flat-packed furniture allows more furniture to be loaded onto lorries etc., meaning fewer journeys have to be made to deliver the product);
- recycling products at the end of their useful life to ensure they don't end up in landfill.

Designers and consumers therefore have a responsibility to ensure that the products they design and use do not impact on our society and environment. If we do not consider the issues outlined above, we will only further damage our environment and create more problems for our societies across the world.

ONLINE TEST

Head to www.brightredbooks.net to test yourself on this topic.

VIDEO LINK

Have a look at the 'Designed for the environment' and 'Sustainable living' clips at www.brightredbooks.net

ONLINE

Learn more about the environmental and social impact of designing electrical products by exploring the link at www.brightredbooks.net

THINGS TO DO AND THINK ABOUT

Outline the key issues affecting our environment and society with the advancements in technology and product design.

FUNCTION, PERFORMANCE AND FITNESS FOR PURPOSE 1

WHAT IS FUNCTION?

Function is described as the job a product is designed to do. All products have a purpose, but many have more than just one job. When discussing function, we often refer to a product's primary and secondary functions. You must learn the difference between primary and secondary functions and be able to explain this.

Primary and secondary functions

The **primary function** of a product is the main job a product is designed to do. **Secondary functions**, on the other hand, are described as other functions the product can perform. Secondary functions can also be in a product's styling and appearance. Some secondary functions are planned by the designer, and others come about because users have used a product in a way that was unintended by the designer.

Intended secondary functions: The primary function of the desk shown is to provide a working space for the user. The secondary function of the desk is storage. Without the secondary function, the desk would still be able to perform its primary function effectively.

Unintended secondary functions: A chair's primary function is to sit on. An unintended secondary function would be someone using the chair to stand on while changing a light bulb or attempting to reach a shelf.

When designing a product, it is important that the designer focuses quickly on the primary function. Taking the example of the desk, a designer could quickly design several different ideas based on its primary function without the need to include any secondary functions. After some initial designs have been developed, the designer could then begin looking at secondary functions such as storage to further enhance the product. Designers also have to take into account that consumers may not always use the product as was originally intended. Therefore, in the chair example, designers must fully consider other issues such as strength and stability in relation to function to ensure the product is safe even when misused.

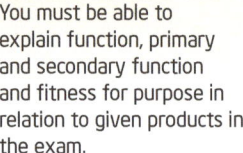

DON'T FORGET

You must be able to explain function, primary and secondary function and fitness for purpose in relation to given products in the exam.

ONLINE

Read more about functional design at www.brightredbooks.net

VIDEO LINK

Have a look at the 'Human Car' clip at www.brightredbooks.net. What do you think the primary and secondary functions of this product are? How do you think it lives up to its function?

ONLINE TEST

Test yourself on function at www.brightredbooks.net

WHAT IS PERFORMANCE?

Performance describes how well a product performs its primary function. Consumers often want products that have a good life expectancy and perform their jobs well. However, a product's success in relation to performance can differ based on its cost and value for money. Cheap sunglasses, for example, may be purchased for a holiday where the user only requires the product for a short period of time. We would not expect these to last as long as a designer pair, to be as comfortable to wear or to perform as well in terms of blocking out sunlight. Various aspects affect a product's performance, and the designer must consider these.

Durability

Durability is defined as a product's ability to withstand wear, pressure and damage. In an ideal world, we would like all products to be durable so that they have a high life expectancy and are safe to use. This cannot always be the case, however. Factors such as budget, materials, construction, who will use the product (market) and how they will use it, all determine a product's durability performance. A product that is designed to be inexpensive and aimed at a budget market cannot make use of the best components, materials and construction processes available. This means the product will, potentially, not last as long. Some consumers misuse products, as discussed previously, and this can cause them to degrade quicker and breakdown or become unsafe. Designers therefore have to balance all of these issues when designing products to ensure they remain as durable as possible and perform as expected by the consumer, based on the price paid.

Materials and construction

Materials and construction greatly affect durability and performance. Understanding how the product will be used and where it will be used will help us to determine:

- the **material** properties required, such as strength, weight and resistance to shock, water and heat

- the **construction** techniques required in ensuring that a product is long-lasting, hard-wearing and sturdy.

For example, a mountain bike which is designed to be used off road and subjected to rougher terrain, adverse weather, jumps and impact cycling will require materials that are hard-wearing, non-corrosive and shock-absorbent. The construction used will also have to be suitable to stop the bike from becoming damaged or structurally weakened during extreme use. If the bike doesn't perform in these areas, this can limit the consumer's confidence in using it and potentially make the product unsafe.

Downhill mountain biking

THINGS TO DO AND THINK ABOUT

Find some everyday products in your home. Evaluate how well these products perform, based on the key performance factors outlined on pages 23 and 24. Try to establish where improvements can be made. This will help you in preparing for exam questions based on performance.

FUNCTION, PERFORMANCE AND FITNESS FOR PURPOSE 2

PERFORMANCE (CONTINUED)

Ease of use and size

No one wants to buy a product that is difficult to use. Poorly designed products can cause us a great deal of frustration and can in effect make our daily lives more difficult. It is therefore important that the designer considers the product's functionality in relation to its intended user, to ensure that it is fit for purpose and performs its job well. Can-openers, for example, come in a variety of designs. This is due to the fact that a standard can-opener will not suit the needs of everyone. Older people, who may have less strength than younger people, will find it difficult to turn and grip the can-opener, as will users with medical conditions such as arthritis. Some designs now include spring-loaded handles and automatic cutters to make this task easier.

 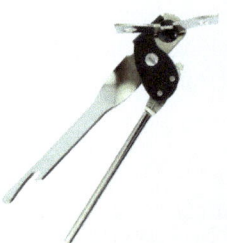 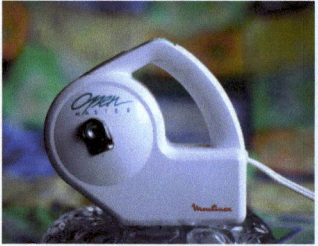

Different can-openers

Miniaturised digital camera

Size is another aspect that can affect ease of use and performance. Electronic products such as digital cameras are becoming smaller with the continual advances in micro technology. Smaller products mean that parts such as buttons also become smaller. This creates an issue for the designer, as smaller buttons and key parts that require the user to operate a product can make it more difficult to use.

Ease of use is also a key focus of **Ergonomics** (pages 34–35) and you must think about and link this when considering how ergonomics influences a design.

Ease of maintenance

Products are normally maintained or repaired to keep them functioning as they should do.

This can involve:

- running costs such as batteries
- repair costs for replacing broken parts
- general maintenance such as cleaning.

If a product can be easily maintained, then it can often outlive its expected life, delivering **value for money** where the cost of the maintenance is less than buying a new product. With bicycles for example, consumers know that with prolonged use, parts such as tyres, chains, brakes and gears will need to be replaced, cleaned or oiled. Carrying out these tasks should be easy, and the cost of doing so should be relatively cheap. However, designers do not always consider ease of maintenance to be of high importance. Products like mobile phones and digital cameras are not made easily maintainable by the consumer. These products normally require people with specialist skills and tools to maintain and repair them when they become damaged. This can result in high costs – and sometimes the repairs can cost as much as buying a new product. In the electronics market, where technology moves so fast and new products are being developed all the time, functionality is more important than maintenance, as most consumers will try to keep up to date with the latest trends in the market.

FITNESS FOR PURPOSE

Another important aspect of **function** and **performance** is fitness for purpose. Fitness for purpose describes how well a product carries out its intended job. It is obvious that a product should carry out its job; but how **well** should a product do its job? Some products do not perform as well as intended due to poor design, and this can also depend on whether or not the product is being used in the correct situation.

Handheld vacuum cleaner

Case study

The primary function of a vacuum cleaner is to remove dirt, and its secondary functions are to be easily emptied and easily manoeuvred. Look at the three vacuum cleaners shown. If we consider the primary and secondary functions, each of the vacuum cleaners performs well. However, if we swap each vacuum cleaner's recommended place of use, we will quickly find that each product no longer performs in the way it should do.

Handheld vacuum cleaner: This is used for small jobs such as vacuuming the seats of a car. If we tried to clean an entire house with this device, it would no longer be fit for purpose, as it would not be able to hold as much dirt as the household vacuum cleaner. It would take far too long to clean an entire house with such a small product.

Household vacuum cleaner

Household vacuum cleaner: If we were to try vacuum-cleaning an industrial site containing large amounts of debris, then the household machine would not be fit for purpose. A household vacuum cleaner is designed for lifting general domestic dirt and would not be able to cope with lifting large debris, which could damage its workings.

Industrial vacuum cleaner

Industrial vacuum cleaner: If we were to use the industrial vacuum cleaner in our homes, it would no longer be fit for purpose. It would be awkward to manoeuvre around the house and to take upstairs due to its size. Also, the power of suction from the vacuum could lift and damage carpets, making it impractical.

Product lifespan and the environment

A product's lifespan is largely determined by all of the design issues outlined in this chapter. Products that function/perform poorly in these areas and those that are not fit for purpose are likely to be discarded more quickly, meaning they will end up in landfills. In addition to this, **obsolescence** will also determine a product's life span. For example, products are often discarded for the next 'best thing' because of:

- **ordinary obsolescence**, where products naturally go out of fashion or become obsolete due to changes in consumer demand and new technologies. Or

- **planned obsolescence**, where designers/companies design products to wear and break down over a period of time, so that new products which utilise new manufacturing technologies, styling and technologies can be introduced to the market.

With this in mind, it is now more important than ever that designers and manufacturers consider the implications of the decisions they make when designing products. We, as consumers, also have a responsibility to consider how we use and recycle products at the end of their useful life. If we continue to fail to design, use and recycle products effectively, we are at risk of continuing to damage and destroy our environment.

THINGS TO DO AND THINK ABOUT

Choose a simple everyday product that you feel is not fit for purpose, and write a short paragraph describing why. Discuss this with your teacher and class.

DON'T FORGET

Performance is affected by many factors. You should ensure you understand these and can discuss them when asked about specific products. Not every issue of performance will be important to every product.

VIDEO LINK

Watch the video on performance bicycle clothing, and consider how performance has influenced the design of the product, at www.brightredbooks.net

ONLINE

Follow the link at www.brightredbooks.net for more on this.

ONLINE TEST

Head to www.brightredbooks.net to test yourself on this topic.

MARKET 1

WHAT IS MARKET?

A market can be described as any place where businesses and consumers trade products. Markets constantly change, and businesses must be aware of current market trends as well as of the changing needs and wants of consumers. This keeps businesses competitive and allows for the development of new products and markets. There are several aspects of market that you must learn for your National 5 exam.

Simple design that is fun and durable for young children and not expensive for parents to buy

MARKET SEGMENTS

Markets contain all different types of consumers, and designers often try to find ways in which to target these consumers. One way of doing this is via market segmentation. Market segments place consumers into groups of people with similar interests and purchasing decisions. Designers can then find opportunities for product developments within these segments. There are four key market segments, each with its own set of variables:

- **Geographic:** continent, country, city, climate.
- **Demographic:** age, gender, occupation, income, culture, education, religion, ethnicity.
- **Behaviouristic:** readiness to buy, purchase frequency, brand loyalty, benefits sought.
- **Psychographic:** lifestyle, personality, interests, values, social class.

Using these segments, designers can design a range of products to suit a wide variety of consumers. For example, when we consider the **demographic** of **age** and relate this to wristwatch design, we see that the choices of materials, colour schemes, functionality and style all change in line with users' age and the way in which consumers will use the product.

More sophisticated design but still retaining a youthful, fashionable design for teenagers

Very formal and sophisticated, utilising expensive materials that will appeal to adults

THE MARKETING MIX

Marketing mix is often referred to as the four P's of marketing. Bringing a new product to market is not as easy as it may sound. Every day, new ideas and products are being conceived. To bring a new product to market, companies must have a clear vision and plan of what they are doing and be able to do it quickly. Once a company has determined their target market, they can then begin to consider the four P's to help them strategise how they will make a product attractive to the consumer. These are:

- **P**roduct: Any goods that can be sold to the consumer to satisfy demand, wants or needs.
- **P**rice: How much are consumers willing to pay for the product?
- **P**lace: Where is it best to make the product available to the consumer?
- **P**romotion: How can we effectively advertise the product to the consumer?

Companies will often mix these as they see fit in order to create a unique selling point (USP) for the product. The USP will help the product to stand out against its competitors and grab consumer attention.

DON'T FORGET

A **target market** is the group of users a product is aimed at. Target markets can be open or very specific. Various things such as age, gender and lifestyle define a target market.

INTRODUCING NEW PRODUCTS

New products are introduced for several reasons. It may be that the market demands a product to fulfil a certain need or want or to allow companies to remain competitive. It may be that new technologies drive new products which are then pushed onto the consumer. Gaps may open up in the market which companies can exploit. Developing new products is risky and costly, and companies must ensure that there is a market for the product. On the other hand, companies can play safe and aim to be the second developer in the market when a new product is released. This allows them to develop a new product that competes effectively with the original, finding areas where they can improve on the design. Companies can also aim to build upon a previous product's success. For example, the developer of a successful set of headphones can then look to expand their market share by developing new portable speakers and docking stations, and could even collaborate with other companies. This could involve offering the inclusion of your product in new products such as laptops and cars for their audio systems. The company should ensure the strategy chosen is correct for their new product.

HUMAN **NEEDS** VERSUS HUMAN **WANTS**

Needs

A **need** is for something you **have to** have, something you cannot do without. There are many human needs, from the most basic – such as food, water, clothing and shelter – to the more social, such as love, friendship and the need to belong. Designers and companies obviously do not create these needs, as they are part of human biology and psychology. When a person feels that a need is not being fulfilled, they will likely seek a solution or a product that will satisfy it. Having an awareness of this allows the designer to develop and produce new products that will solve this problem, satisfying our needs.

Wants

Wants, unlike needs, are for the things we would **like to** have. They are often described as the products that will satisfy our needs and are desirable rather than essential. Humans often want new products for various reasons:

- to make us feel good
- to have the same as everyone else due to peer pressure and the desire to belong
- to improve our social status
- to make our lives easier.

These wants are important for designers to consider and allow them to continually develop new products that will provide us with satisfaction.

THINGS TO DO AND THINK ABOUT

1. Describe the difference between consumer needs and consumer wants, and explain how designers can use these to develop new products.

2. Consider the products you use every day, and think about how ideas of market segments may have affected their design. Record this in your notes, helping you to apply your learning.

ONLINE TEST

Test yourself on markets at www.brightredbooks.net

VIDEO LINK

Watch the video clips at www.brightredbooks.net. These videos will help you to further understand the concepts of marketing.

ONLINE

Learn more about market segmentation and the marketing mix at www.brightredbooks.net

MARKET 2

MARKET PULL VS TECHNOLOGY PUSH (MARKET PUSH)

Market pull and technology push are two of the main driving forces for new product development. Market pull is where demand on the market is pulling designers towards designing new products, and technology push (market push) is where new product developments are pushed upon us, creating a demand for them.

Market pull

Market pull happens when a demand is created for a product to fulfil a need or want. A good example of this is the development of filmless cameras. The first portable cameras required a film to capture photographs, which then had to be developed before the photograph could be produced. This was a time-consuming process for the consumer, and the market demanded a more efficient way to capture and produce photographs. This eventually led to the development of the digital camera. This sort of demand forces designers and manufacturers to continually seek to improve products for their market, satisfying the consumer.

Technology push (market push)

Technology push can be achieved through **product evolution**, where existing products are continually updated and improved to create demand for the latest version. Alternatively, a new product can be developed that hasn't been seen before due to scientific research, advances in technology and new materials and advances in manufacturing processes. Designers can then use these developments to create new products. Once a product has been developed, it is then combined with a high-quality marketing campaign to create demand for the product, pushing it onto the consumer. This is a risky strategy, however, as product development costs a lot of money. For example:

The development of miniature electronics has revolutionised the size of our electronic devices such as phones, cameras, televisions and computers. Touch-screen technology has also revolutionised the way we use these products. Smart watches, touch-screen TVs, tablet computers and e-readers that were once new products are now commonplace in society.

The development of smart eyeglasses, however, is an example of technology push that hasn't quite worked. Although these products have utilised good marketing campaigns, consumers and society have not seen the need for such a product, and this has stunted the product's development and growth, costing companies money and time.

CONSUMER DEMAND

Consumer demand is linked directly to product popularity and drives **market pull**. Once we have become fixed on the need for a product or on a new type of product, we will demand more of it. Imagine life without smartphones or computers. Manufacturers therefore mass-produce a product to ensure there is enough of it to satisfy demand. Due to this market pull, designers continually develop products. Consumer demand also affects products that wear out quickly, such as fashion wear, cleaning products and car tyres. These products come with high demand because they require continual replacement.

SOCIAL EXPECTATIONS

Consumers now, more than ever, are aware of design and how it affects us – so much so that we have now created expectations of what specific products should be like. For example, we expect modern mobile phones (i.e. smartphones) to do more than just make calls and send texts. Designers therefore cannot simply design a similar smartphone or one with fewer features than the previous model. Due to the continual development of products and the social expectations placed upon them, we expect new products to:

- perform better
- have better aesthetics
- be more environmentally friendly
- represent better value for money
- give us more choice
- come with newer technologies and more features.

The designer must therefore consider market research to ensure that new products meet consumer expectations.

MARKET NICHE

Niche marketing describes the process of targeting specific groups of people within a market segment and finding a product to suit their particular needs. Products developed through niche marketing will not readily appeal to the wider marketplace and will instead focus on a profitable area of the market. If a company identifies a niche market, it can capitalise on this – and quickly, gaining a majority hold over the market, where no one else will have developed the product yet. A good example of this is camping stoves. Camping is a specific market, and the need for a small portable camping stove is something consumers in this market demand. Various designs have been created to capture this market niche.

Various stove designs

BRANDING

Branding tells us who a company is and what they are about. It plays a crucial part in influencing our purchasing decisions and our expectations of a product. A company's name and logo are two of its biggest assets, as they allow consumers to identify the brand. Branding can influence the consumer in the following ways:

- It creates loyalty to a product, where we believe in it and define it as a reliable, successful and good brand.
- It makes us feel part of something, as our family and friends may have the same brands, or because it is a popular choice among the majority of consumers.
- It can satisfy our needs and wants.
- It can affect how we see a product. We associate brands that we like with quality regardless of other people's perceptions.

DON'T FORGET

Make sure you can explain the terms listed and know how they influence the design of products.

CAPTURING CONSUMERS THROUGH ADVERTISING

Promotion of a product is vital to its success. Products are advertised in various ways to heighten interest in the product and to make us want it. Companies can do this in the following ways:

- TV and internet adverts
- promotional adverts such as billboards, magazines and posters
- celebrity endorsement of a product
- special discounts, such as discounts for buying a product within a certain time, or 'buy one, get the other half-price'
- pairing products with other deals: for example, buy a new smart tablet and get 10% off online music/video services
- offering vouchers online or in store for this or future purchases.

VIDEO LINK

Learn more about branding by watching the clip at www.brightredbooks.net

THINGS TO DO AND THINK ABOUT

Consider making your own revision notes to help you condense these topics into your own words. For each of the above topics, you should write a short definition, as this will help you with explaining these terms in your final exam.

Furthermore, you should look at common products found around the home and think about the market considerations that were made when developing these products. You could discuss this with friends in class or with your teacher.

ONLINE TEST

Test yourself on market at www.brightredbooks.net

AESTHETICS 1

WHAT ARE AESTHETICS?

DON'T FORGET

When answering questions on aesthetics, it is important that you refer to the key aspects of aesthetics to help you to attain full marks.

Aesthetics can be easily described as the appearance and styling of products. It influences our purchasing decisions and can make a product's function seem obvious or confusing. What people likes will differ from person to person. Designers often look at markets to determine what individual groups like and to assess what fashions or styles are popular at present so that they can design products that appeal to us. There are many aspects of aesthetics that influence design, and you will have to learn each of these.

COLOUR

Colour is more than just the colours we see every day. The choice of colour is very important, as getting it wrong can affect how well a product will sell. This is down to the fact that colour can enhance a product's shape and form, making it stand out and appeal to us. Colour also relates to psychology and can evoke emotions and tell us things about a product. For example, by colour-coding switches in machinery such as green for 'on' and red for 'off', it makes it easier for us to understand how to use the product. Red also highlights the emergency stop in machinery and alerts us to the safety aspects of the product. Colours also influence certain emotions – and designers can use these to create an emotional connection with a product. These emotions link directly with psychology in ergonomics (see pp. 34–35) and can help us to better understand a product's function. For example, we understand that a green switch turns something on and a red switch turns something off or represents an emergency stop or danger.

Colour	Emotions/connections created
Red	Warmth, anger, love, passion and danger
Yellow	Warmth, happiness, bright, uplifting, positivity – and alerts us to hazards
Orange	Warmth, bright, happiness and bold
Blue	Cold, calm, loyalty, peace and sad
Green	Cool, natural, safety, envy and environmental
Purple	Cool, royal, luxurious, wealth and sophistication
Black	Stylish, timeless, sleek
White	Hygienic, purity, timeless and innocence

SHAPE AND FORM

The simplest way to understand shape and form is that **shape** refers to a flat two-dimensional shape, and **form** is the three-dimensional composition of that 2D shape – e.g. a 2D square becomes a 3D cube. Shape and form are very important, as they define the look of a product and how we visualise it. Shape and form are often thought of as being geometric or organic.

Geometric: precise and mathematical in composition, consisting of 2D shapes such as squares, triangles and circles, and 3D forms such as cubes, prisms and cylinders.

Organic: inspired by the natural world, and free-form in composition. This could be likened to the curvature of a wave.

Both of the above can provide us with very different aesthetics in product design. Look at the two bookshelves shown on the next page.

contd

The geometric design is very structured and its function easy to identify. The organic design is based on a tree and makes for a unique piece. This also changes the functionality of the shelving as the distribution of space becomes harder to utilise. However, both products can still be considered attractive. Shape and form can therefore be used to create interest in a product and to help us better understand its function, and can allow designers to create designs that make us question what we believe a given product should look like.

Organic bookshelf

PROPORTION AND SIZE

Proportion describes how well the differing parts of a product relate to one another in terms of size. When all parts of a product are unified, they create a harmonious design that conforms with our understanding of general everyday products. When thinking about proportion, we often consider visual balance in terms of **stability**, **symmetry** and **asymmetry**.

For example, the chair shown here looks stable due to its symmetry and structured design. It is well balanced with all of its features working well together creating a pleasing design.

The following chair looks unusual as common chairs would likely have zero or two arms. The asymmetrical seat design causes us to question the seat's functionality. Furthermore, the thin legs seem disproportionate when compared to the thickness and size of the seat, which could make us assume the chair may lack suitable strength. This shows us that experimenting with proportion can make unique products and create interest. However, when we make items look disproportionate, people can often form negative opinions of the product.

ONLINE TEST

Test yourself on function at www.brightredbooks.net

ONLINE

Learn more about aesthetics at www.brightredbooks.net

 THINGS TO DO AND THINK ABOUT

Consider the shape, form and proportions of this bookshelf. Write down your thoughts in your revision notes.

AESTHETICS 2

TEXTURE AND MATERIALS

After we look at a product, our next instinct is to touch it. **Texture** considers the feel of a product's materials and finish when touched. Texture can make products feel cheap or expensive. It can also create fun and can heighten our enjoyment of using a product because we like how it feels. Texture can also be used as a safety feature to provide grip. For example, the rubber grip and pattern on this hammer handle makes it easier to hold and more comfortable for us to use.

Different **materials** provide us with different textures, and different **finishes** can be used to enhance a material's appearance and texture. Visual texture can also be used on a product to enhance its design. For example, these kitchen drawer fronts are made from low-cost chipboard. However, the printed hardwood plastic laminate applied on top makes it look more luxurious with its hardwood aesthetic.

CONTRAST AND HARMONY

Contrast can be described as things that oppose each other, and **harmony** as things that work well together. Both contrast and harmony can be relevant in other aspects of aesthetics.

Colour: Warm colours such as red, yellow and orange will all harmonise, as will the cool colours green, blue and violet. However, warm and cool are a contrast, and therefore warm and cool colours will oppose each other, e.g. yellow and purple.

Shape: Geometric shapes will work well together but will contrast with organic shapes.

Materials: Wood and metal contrast, as do wood and plastic. This contrast can create unique products and can also make products seem stylish if done well.

Designers choose to use harmony and contrast for different reasons. Harmony can make a product seem more pleasing to the eye, whereas contrast can make product seem bold or exciting. Contrast can also be dangerous, however, in making some products look unpleasant. The chairs shown here provide a good example of this.

The flowing organic design of this chair, along with the use of one main material, creates a harmonious design and looks pleasing to the eye.

The bold geometric shapes harmonise well, whereas the contrasting warm and cool colours make the product stand out.

FASHION, STYLE AND FADS

Fashion is something that is current and popular among consumers. It will remain fashionable for a period of time until the next new fashion comes along.

Style is distinctive and is something that lasts long after its peak period. For example, there are many architectural styles that still influence design today, such as Bauhaus and Art Deco. It can be adapted when necessary and can be used to influence the aesthetics of products.

A **fad** is short-lived in comparison to fashion and style. It is something that becomes popular very quickly but then disappears after a short time.

Knowing what is current among consumers can be important in creating a product that identifies with them instantly. Designers often look at current fashions as well as existing styles in helping to create aesthetically pleasing products. It is important to know what fashion, style and fads are, and how these can be used when designing products.

CASE STUDY

In the exam, you will be asked to consider the aesthetics of various products. To help you prepare for this, we will consider the aesthetics used in the design of this drill.

The **colour** scheme uses a bold black-and-yellow design that would normally alert us to a hazard. However, in this case it forms an industrial-looking design, has strong brand identity, fits the drill's environment and makes it easily seen. Notice also that all of the parts we would operate are black, helping us in understanding how to use the drill (this relates to psychology in ergonomics).

The **shape** is geometric, again relating to the structural nature of the product's environment. The handle and button shape consider the contours of the hand and shape of the fingers. This makes it easy to identify where to hold the drill and what to press.

The asymmetrical **proportion** of the drill allows us to easily identify the front and back. The drill is well balanced proportionally, and the base of the drill looks stable.

The **textured** grip on the handle makes it easy for us to identify where to the hold the drill. It also makes the drill look robust and *safe* to operate, reassuring us.

Being able to evaluate a product's aesthetics in this way will help you in completing your exam and coursework.

 ## THINGS TO DO AND THINK ABOUT

To help you prepare for your exams, you should begin evaluating products, focusing on aesthetics. Being asked to describe a product's aesthetics, or the issues a designer would have to consider based on aesthetics, is a common question that will come up in the exam. You should record your work in your own revision notes and could discuss your findings with a classmate.

 DON'T FORGET

Colour, shape, harmony, contrast and materials are five of the easiest aspects to discuss when focusing on aesthetics.

 ONLINE

Read more about aesthetics at www.brightredbooks.net

 VIDEO LINK

Have a look at the 'Visual Factors' clip at www.brightredbooks.net

 ONLINE TEST

Test yourself on function at www.brightredbooks.net

ERGONOMICS

WHAT IS ERGONOMICS?

Ergonomics is the study of how humans interact with everyday products. The majority of consumers don't think about good design, unless it is exceptional, because we have no cause to. However, we notice poor design. Think of the number of times you have tried to use a product and found it difficult. Poorly designed products inconvenience us and can even lead to injury. Poor design will discourage us from using a product and force us to seek an alternative. Designers therefore need to consider ergonomics so products suit human needs. If a product is designed around the systems of human functionality, we can enjoy the experience of using it, and it can make our daily lives easier.

When considering ergonomics, the designer must look at three key areas:

- anthropometrics
- physiology
- psychology.

ANTHROPOMETRICS

People all over the world come in different shapes and sizes, defined by various factors like *age*, *gender* and *ethnicity*. So, designers must use **anthropometrics** to help group similar types of people together to fully understand the dimensions of the human body and how to design products to suit the needs of different groups.

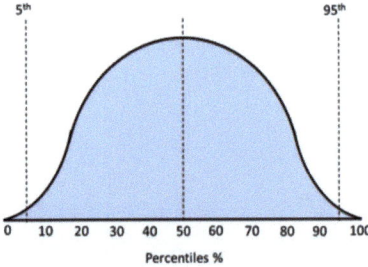

Percentiles bell curve

Designers often use percentile distribution to describe the range of body dimensions for a given group of people. In any group, the majority of users fall between the 5th and 95th percentiles, which includes the average-size at the 50th percentile. If we refer to stature, the 5th percentile is the data for people in the shorter range within the group, the 50th percentile is the data for the average height, and the 95th is the data for people in the taller range. Percentiles allow designers to quickly establish what sizes would be the most important to use for a given product. For example:

- A door opening would have to suit the 95th percentile of people because, if the majority of tallest and widest people can fit through it, then so can the smallest.

- A wall cupboard would have to suit the 5th percentile because, if the majority of smallest people can reach it, then so can the tallest.

However, we mustn't forget the 0–5th and the 95–100th percentiles. People in these ranges will be outwith common percentiles and will each make up 5% of the population. These people often require specialist products designed to satisfy their needs.

Once a percentile group has been established, a designer can then begin to focus on critical sizes to determine what size a product and its individual elements should be.

Determining critical sizes

If we were designing a chair and determining what width the seat should be, we would consider hip breadth. An anthropometric data table would help us determine the required size.

	User group to be considered					
	UK adult male			UK adult female		
	5th	50th	95th	5th	50th	95th
Hip breadth (mm)	310	360	410	305	370	435

From this data, we can establish that the 95th percentile size for females gives our required width. If we design the chair to be no less than 435 mm wide, then the majority of users could sit on it comfortably.

PHYSIOLOGY

Physiology is the area of ergonomics that deals with the physical capabilities of the human body. This can involve looking at strength, posture, movement, flexibility, reaction speed and muscle control. The data collected in this study will define any constraints a designer has to apply to make any given product easy to use. For example, think about a standard games controller. How much effort must a user apply to press its buttons? How heavy should it be so that it is easily lifted and balanced in the hand? Should there be finger indents, curvatures or materials that make it easier to grip and hold? If any of these issues are overlooked, then users may encounter several problems when using the product.

We should also consider **percentiles** in relation to physiology. For example, the effort required to press a button would be focused at the 5th percentile. If the majority of people who are considered to be weaker than average can press the button easily, then everyone else, being stronger, should have no problem pressing it. Considering percentiles will ensure we get the physiological aspects of the design correct.

PSYCHOLOGY

Before choosing to use a product, we look at it first. This allows us to assess whether a product appeals to us. It makes us consider the functionality of the product, and it allows us to assess any safety issues that the product may present. If a product looks good, easy to use, safe or interesting, we will be more likely to interact with it. Therefore, it is very important that designers consider human psychology when designing products. Designers must know how we will perceive a product and process information when using it to ensure that we understand what to do when operating it. Thorough consideration of this allows users to operate the product correctly and safely, while also saving us from becoming frustrated by the product if we can't work out how to use it.

When considering psychology, we should think about how colour, symbols/images, sounds and our sense of touch will help us to understand the product better. If you consider a power switch, the fact that it lights up green tells us that it is on. The colour green also represents 'go' or 'on'. The power symbol is a common symbol that we understand as a power switch. The button may also make a clicking noise when pressed, or we may feel a click when we press it. These psychological aspects of design are extremely important when considering how we understand and interact with products.

DON'T FORGET

When discussing anthropometrics and physiology, try to refer to user percentiles. When answering ergonomic questions specific to a given product, you should always try to refer to all three ergonomic areas to ensure you gain full marks.

ONLINE TEST

Head to www. brightredbooks.net to test yourself on this topic.

THE ERGONOMIC STUDY OF AN EVERYDAY KETTLE

So that you can better understand how ergonomics works, an ergonomic evaluation of a standard domestic kettle is given below.

EXAMPLE: Standard electric kettle

Anthropometrics:

Size to be determined	Percentile range	Reason
1. Diameter of the handle	5th–50th	If the smallest hand can comfortably grip the kettle, then so can the largest.
2. Gap between the handle and kettle	95th	If the largest hand can fit through the gap comfortably, then so can the smallest.
3. Size of the switches	50th	If the buttons are designed for the average finger width, then the majority of users could press them comfortably, as they will not be too big or too small.
4. Position of the lid switch on the handle	5th	The switch that operates the lid cannot be too far from the user's thumb when gripping the handle, as this may require the user to overstretch.

Physiology:

- **Weight:** 5th percentile. The kettle must be easy to lift and tip in relation to its weight when empty or full.

- **Grip:** The materials and shape of the handle must be easy to grip in relation to weight, to limit the possibility of dropping the kettle.

- **Pressing switches:** 5th percentile. The user would expect this to be easy, and therefore they should not require significant force to operate the kettle.

Psychology:

- **Colour:** The parts the user must operate are black in colour. This makes it easier for the user to identify how to use the product.

- **Sound:** When switches are pressed, they make a clicking sound. This noise tells the user that something has been switched on.

- **Light:** The filling indicator turns blue when the kettle is on, indicating that the kettle is in use. Some kettles do this in even more intuitive ways where the colour of the light changes from blue to red as the water temperature increases.

- **Symbols:** The filling indicator has a key that tells us how many cups can be made in relation to the water level. This saves the user from over- or under-filling the kettle, which could lead to waste or damage.

Ergonomics plays a vital part in design. Without it, products would be difficult for us to understand and use.

THINGS TO DO AND THINK ABOUT

Find a range of products around your home. Study them, and then evaluate their ergonomic design. Think about what has been covered above, and see if you can identify key ergonomic features. It would be useful to bring your findings to class and discuss this.

ONLINE

Learn more about ergonomics by exploring the link at www.brightredbooks.net

EVALUATION TECHNIQUES

EVALUATION TECHNIQUES

The following evaluation techniques can be used throughout the design process, helping the design team to find product opportunities **or** to evaluate their solutions. This is extremely beneficial in testing solutions for refinement or in confirming the solution is suitable for its intended purpose.

Questionnaires/surveys

As described previously, on page 12, questionnaires are a quick and effective way to canvas consumer opinion. In the same way, they can be used to find potential markets they can also be used to help you evaluate your solution.

Focus groups

A focus group is a form of research in which a group of people are asked about their opinions on a given product. The group of people chosen could be specific to a target market or members of the general public. Questions are asked to prompt discussion about the product, and the group's answers are recorded. A focus group allows people to get hands-on with the product, which often stimulates more effective discussion when evaluating **function** and **aesthetics**.

User trials

Different from 'user trips', user trials require the consumer to test a product. This is an excellent method for evaluating **function** and **ergonomics**. The person completing the user trial is usually given a series of tasks to complete. For example, if we were testing a kettle, the user may be asked to evaluate:

- How easy it is to understand and use?
- How easy it is to lift and move?
- How easy are the buttons to press?
- How comfortable is the handle to grip?

The results would then be recorded and conclusions drawn.

Test rigs (product testing)

A test rig is a way of physically testing the **performance**, extremities and limitations of a product's **function, durability, materials** and **safety**. The product will be put through a series of tests that aim to provide results based on the aforementioned design factors. Using the same example, if we performed a test rig on the kettle, we may test:

Testing a chair using high-pressure pneumatics to assess its strength under different weights

- How quickly it can boil water?
- How do the materials perform if the kettle is left boiling continually without switching off?
- Do the materials break easily if the kettle is dropped?

contd

Measuring and recording

This is an effective method for evaluating **function**, **ergonomics** and **performance**. For example:

- **Function**: If designing a bookshelf, we could measure and record a range of book sizes, allowing us to work out the best possible dimensions for shelving space.

- **Ergonomics**: Measuring and recording allows us to effectively measure the dimensions of the human body, which are crucial in ensuring that products are suited for their human users. See 'Ergonomics' (pp. 34–35).

- **Performance**: In combination with a test rig, we could measure and record aspects such as sound and weight in portable speakers, allowing us to establish how well the product performs.

Online searches

Online searches using the internet can be useful in helping to find information about competitor products and items such as consumer reviews of products. Images and data can then be collated, and conclusions can then be drawn from them.

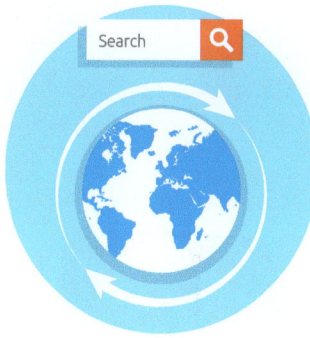

Product comparison

A product comparison compares similar products within a similar market area and price range. This can be useful for evaluating many factors such as **function**, product features and **aesthetics** to establish what is currently on the market and what our competitors are doing. However, it is particularly effective in evaluating **value for money** and **cost**. For example, if we compare toasters, we could conduct the following evaluation:

Features	Product 1	Product 2	Product 3

Colours range	Black/silver	Red/black	Gold/white
Speed of toasting	2 minutes	1·5 minutes	1 minute
Slices held	2	2	3
Product guarantee	1 year	1 year	2 years
Cost	£20	£40	£32

From the comparison, we can conclude that toaster 3 is the best option based on cost and value for money, as it clearly has more attractive features at a lower price. It is important, however, that we do not compare products designed for different markets.

Comparison to the specification

One of the most assured methods of evaluating your product's suitability for its intended markets, is to compare it back to the specification. This should happen continuously throughout the design process and will ensure that your final design meets the requirements as it set out in the original specification.

 ONLINE TEST

Test your knowledge of evaluation techniques at www.brightredbooks.net

 THINGS TO DO AND THINK ABOUT

During the course of National 5 you will have to learn to evaluate products and design factors for developing items such as specifications. You should conduct some of the evaluation techniques outlined above to help you do this effectively and further develop your understanding of these.

MATERIALS

SELECTING MATERIALS

There are three main categories of material that you must learn about: **wood**, **metal** and **plastic**. This chapter will explain the importance of materials in manufacture and the considerations we must make when using them.

WHAT MATERIAL?

During the ideas stage of the design process, product designers attempt to be as innovative and creative as possible. As designers develop these ideas towards manufacture, they must consider how an item will be manufactured with specific regard to material choices and manufacturing processes. Different materials offer designers all sorts of properties that can affect the aesthetics, cost, production and functionality of items. When selecting a material, designer and manufacturers must consider the following:

- Is the material readily available?
- What properties should the material possess?
- Where will the product be used? How will this affect the choice of material?
- What is the environmental impact of using this material?

These teak planks come in a range of standard lengths, widths and thicknesses

MATERIAL AVAILABILITY, STANDARD LENGTHS, SIZES AND COMPONENTS

Readily available is a term used to describe how easily accessible a material is and if there is a sufficient stock of it. Materials that are readily available will be cheaper and will allow manufacturing to begin more quickly. However, where materials are not readily available, order times can slow the manufacturing process and can also raise costs.

Standard lengths and sizes refer to materials or parts that are easily available and cut to a specific size. For example, wooden manufactured boards can come in standard sizes of 2400 mm by 1200 mm, or 1200 mm by 600 mm. They also come in different thicknesses such as 12 mm, 16 mm and 22 mm. Metal and plastic sheets also come in readily available sizes similar to these. Solid timber, metal and plastic bars/rods also come in pre-cut lengths and thicknesses. These standard lengths and sizes reduce the cost of manufacture, as materials do not have to be cut to a specific size. Manufacturers can also calculate exactly what they need, working with the given sizes to reduce both cost and waste. This also makes it easy to order materials for a particular job.

Standard components are common parts that are used in products such as screws, brackets, wheels, handles, circuits and so on. Using standard components makes design and manufacture easier. It also benefits the consumer, as replacement parts can be bought when items break.

Standard components

DON'T FORGET ✚

You may be asked to explain standard lengths, sizes or components in the exam and the benefits of using them.

MATERIAL PROPERTIES

All materials have properties that ultimately determine their suitability for use in a particular product. Some materials are chosen because of their aesthetic qualities; however, this alone is not enough to determine a material's suitability. Many other properties must be considered. Although you do not need to learn each of these for the exam, it is worth having a working knowledge of them to further enhance your understanding of materials and how they can be used. Here are some key properties to consider:

PROPERTY	DESCRIPTION
Strength	The maximum force a material can withstand before breaking when pulled apart, crushed or twisted
Ductility	The length to which a material can be stretched before breaking
Hardness	How difficult it is to cut or mark a material
Toughness	The amount of energy a material can absorb before it breaks when hit with something like a hammer
Malleability	The amount of shaping that can be done with a material before it breaks in terms of bending, twisting and so on
Brittleness	The material will break very easily when any stress or pressure is applied to it
Elasticity	The length to which a material can be stretched and still return to its original shape
Conductivity	How well a material conducts heat or electricity
Corrosion	Whether a material corrodes (rots, weakens, rusts etc.) easily due to oxidation or exposure to chemicals
Workability	How easy a material is to cut, shape and form

ENVIRONMENTAL IMPACT

We should also consider a material's practicability in terms of where it will be used and its impact on the environment. For example:

- **Is the choice of material suitable for its intended use and place of use?** Metals that rust would be no good in environments where they are constantly exposed to water. This could result in them corroding if not properly maintained, which means they will eventually be discarded.

- **Is the material sustainable or renewable?** For example, can trees be replanted and grown quickly to reduce the impact of logging for timber and deforestation on the environment? Softwoods are a good example of this when compared with hardwoods. Metals, on the other hand, cannot be replaced once mined from the earth.

- **Can the material be recycled or reused?** Recycling materials means that products do not end up in landfill at the end of their useful life. However, recycling alone creates by-products such as gases and chemicals that can be harmful to the environment. Also, some materials can only be recycled so many times before they become unusable. Upcycling, where we reuse materials and components that are still in good condition, is a much better way to reduce environmental impact. Glass bottles are a good example of this, as they can be cleaned and reused without being wasted.

If products are designed without any consideration of the materials we use, then we risk further damaging our environment, where products are simply discarded at the end of their life. Taking the time to look at alternatives is therefore very important in ensuring that materials can be replaced and reused as much as much as possible, to reduce the impact of waste.

VIDEO LINK

Watch the video clip at www.brightredbooks.net for a good insight into how designers select materials for their products.

ONLINE TEST

Test yourself on this topic at www.brightredbooks.net

THINGS TO DO AND THINK ABOUT

1. In your own words, explain the term 'standard lengths' and 'standard components'.

2. Describe the advantages offered by using standard sizes and components.

3. Describe the difference between 'recycling' and 'upcycling' materials.

WOOD

WHY CHOOSE WOOD?

Wood is a natural material that is extremely useful. Wood comes from trees, and each species of tree gives the processed timber its own unique character. It is a traditional building material, and offers designers and manufacturers many **advantages** such as:

● Woods can add natural beauty and character to a product.
● Most woods are easy to work, shape and maintain.
● When used correctly, wood can be very durable, hard-wearing and strong.
● Wood can be environmentally friendly when used from sustainable sources or when recycled/produced and used as manufactured board.

This said, however, there are also many **disadvantages** to using wood, such as:

● Expensive hardwoods are not as sustainable as softwoods due to the length of time they take to grow.
● Using tropical hardwoods also contributes to deforestation of rainforests, which has a negative effect on the environment.
● Solid woods only come in relatively narrow widths, due to the size of tree trunks. This can make the manufacture of large products more time-consuming, as planks often have to be glued together (see 'Laminating', p. 59).
● If not stored correctly, woods can **warp**, altering their shape and making them harder to work – or even unworkable. Woods can warp in several ways as shown on the left.

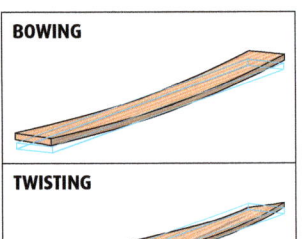

BOWING

TWISTING

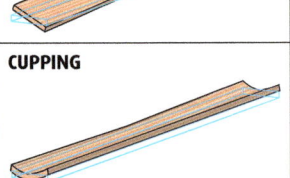

CUPPING

CATEGORIES OF WOOD

Softwoods

Softwoods, such as pine, come from **coniferous** trees. These trees keep their leaves or needles all year round. They can be grown in renewable managed forests, meaning that more trees can be grown to replace what has been used. In general, softwood trees are fast-growing and often lower in cost, and many species can be ready for logging in less than 30 years.

Hardwoods

Hardwoods, such as oak or beech, come from **deciduous** trees. These trees lose their leaves each winter. They tend to grow more slowly than softwoods, taking up to 100 years and more. They are generally more expensive than softwoods; however, they offer better aesthetic qualities. As mentioned previously, the felling of hardwood trees is damaging to the environment and can be a reason for not selecting them in commercial manufacture.

Manufactured boards

Manufactured boards are a much cheaper option than using traditional softwoods and hardwoods. They are produced using wooden particles, strips or sheets. They are often made using recycled materials or by-products such as sawdust from the manufacture of softwoods and hardwood items. Manufactured boards offer the designer several advantages over softwoods and hardwoods:

● available in larger sizes, as they come as boards
● cheaper
● higher stability, as materials like MDF and chipboard do not warp
● **veneers** (thin sheets of softwoods/hardwoods) and plastic **laminates** can be applied to them, allowing
them to simulate higher-quality expensive materials (refer to pp. 46–47).

DON'T FORGET

Learn the descriptions for each category of wood, and make sure you are able to select woods from these when identifying what type of wood would be best for a given product.

DON'T FORGET

You must be able to describe and explain the advantages of manufactured boards over solid woods, the advantages offered by soft and hard woods and the sustainability issues surrounding the use of solid woods.

contd

Softwoods

NAME	PROPERTIES	USES	HOW IT LOOKS IN USE	COST
Scots pine	Straight-grained but knotty, quite strong and easy to work. White/brown in colour.	Low-cost furniture, joinery and construction work.		Low
Red pine	Straight-grained but knotty, quite strong and easy to work. Red/orange in colour.	Building construction. Requires good protection.		Mid-range
Spruce	Quite strong with few knots. Resistant to splitting, but not durable.	Fitted furniture, e.g. kitchen cabinets.		Low
Cedar	Straight-grained and knot-free. Very light and durable. Quite soft.	Construction, sheds and good-quality fencing.		Mid-range

Hardwoods

NAME	PROPERTIES	USES	HOW IT LOOKS IN USE	COST
Oak	Light brown or pinkish brown, hard, tough, heavy and durable. Gets harder with age.	High-quality furniture, garden furniture, flooring, boat construction and veneers.		High
Mahogany	Red/brown in colour, medium weight, quite strong, easy to work and durable, but warps easily.	High-quality furniture, shop fittings, panelling and veneers.		High
Ash	Light in colour, flexible, tough, bends well and varnishes well.	Tool handles, cricket/baseball bats, snooker cues, ladders and veneers.		Mid-range
Beech	Mid-brown colour, hard, strong, tough, tends to warp but bends well. Can be steam-bent.	High-quality furniture, toys, tool handles and veneers.		Mid-range
Teak	Straight or wavy-grained. Coarse uneven texture with oily feel. Golden to darker brown. Very expensive.	Interior and exterior joinery, garden furniture, veneer.		Very high

NAME		PROPERTIES	COMPOSITION	USES
MDF (medium-density fibreboard)		High strength, easily machined, shaped and paints well.	Made from gluing and tightly compressing excess sawdust together.	Furniture, construction and joinery.
Plywood		Very strong, stable and easy to machine/work.	Made by gluing thin sheets together. It is important that the grain of each layer goes in a different direction to ensure maximum strength.	Furniture, construction and joinery.
Blockboard		Very strong, heavy, rigid and good load-bearing capabilities.	Timber strips laid parallel with veneers glued either side with their grain running crosswise.	Quality furniture, worktops, stage flooring and fire doors.
Chipboard		Heavy; and strength is dependent on density of wood chips.	Wood chips mixed with bonding mixture to create large flat boards. Surface and strength depends on chip particle size.	Kitchen cabinets, worktops and carcases for furniture.
Hardboard		Almost like cardboard. Very weak and brittle.	Made from gluing and tightly compressing wood fibres.	Low-cost furniture parts e.g. internal door panels, drawer bottoms and cabinet backs.

Manufactured boards

ONLINE

Learn more about woods by exploring the link at www.brightredbooks.net

VIDEO LINK

Watch the video clips at www.brightredbooks.net to gain an insight into how solid timber and manufacture boards are produced.

THINGS TO DO AND THINK ABOUT

1. Describe the differences between softwoods, hardwoods and manufactured boards.
2. Explain the benefits of using manufactured boards.
3. Using the internet, find images of wooden products, or find wooden products around the home. Try to identify what types of wood have been used to manufacture them based on their grain and colour, and think about why they were chosen. You should record this in your own revision notes.

ONLINE TEST

Head to www.brightredbooks.net to test yourself on this topic.

METAL

WHY CHOOSE METAL?

Metals are extracted from the earth and make up the majority of the earth's elements. Once they have been mined, they are processed into usable materials for manufacture. Metals are processed into a variety of forms including sheets, flat strips, bars, tubes and angular sections. These are all available in a variety of pre-cut sizes, making it easier for manufacturers to work with them, as metal requires a lot of energy to cut and soften.

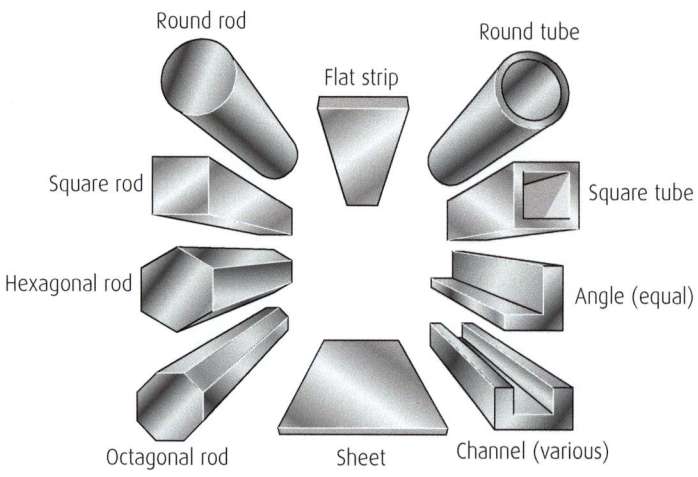

Round rod
Round tube
Flat strip
Square rod
Square tube
Hexagonal rod
Angle (equal)
Octagonal rod
Sheet
Channel (various)

Supply forms of metal

Metals are highly versatile, allowing for flexibility when designing creative and intricate products. Like all materials, metals come with several advantages and disadvantages. Some of these **advantages** are:

- Metals can take a range of finishes from painting, oiling and polishing to plastic dip coating and electroplating.
- They are excellent conductors of heat and electricity, unlike woods and plastics.
- Metals come in a vast range of colours and styles.
- They can be melted, making them easier to mould and form.
- Metals are structurally good, as most are very strong.
- They come in a range of forms suitable for various requirements.
- The majority of metals are easily recyclable.

Disadvantages include:

- Once metals are mined from the earth, they cannot be replaced.
- Ferrous metals rust and corrode.
- Finishes applied to metal will require periodic maintenance, as they can tarnish over time.
- Some metals can be difficult to work with.

CATEGORIES OF METAL

 DON'T FORGET

You must know the difference between ferrous and non-ferrous metals and be able to explain what an alloy is.

Metal is divided into three categories: **ferrous** (containing iron), **non-ferrous** (not containing iron) and **alloys** (combinations of different metals).

Ferrous metals

The term **ferrous** metal refers to any metal that contains iron. Ferrous metals rust over time with continual exposure to moisture or oxygen (oxidation) and are also magnetic due to their iron component. Iron alone, however, has little practical use and is normally combined with carbon elements to create useful metals. These are known as **ferrous alloys**; and some examples are given in the table here.

NAME	PROPERTIES	USES
Cast iron	Relatively inexpensive, brittle and casts well.	Engine parts, machine tools and vices.
Mild steel	Malleable and ductile. Relatively easy to work.	Car body panels, nuts, bolts, screws, tubes, springs and small non-cutting tools.
Stainless steel	Ferrous alloy. Hard and tough. Corrosion- and wear-resistant.	Furniture components, cutlery, sink units, dishes, teapots, boat fittings and kitchen appliances.
High-carbon steel	Malleable and ductile. Can be hardened and tempered.	Cutting tools, files, drills, saws, taps and dies, knives, scribers, lathe tools.

contd

Non-ferrous metals

Non-ferrous metals contain no iron and are referred to as pure metals. You should make yourself familiar with the following non-ferrous metals. Many of these metals are used in applications you will already be familiar with.

Alloys

Alloys are made when two or more metals are combined to create a new material. As mentioned earlier, ferrous alloys are produced to make iron a more practical material. Other metals are often alloyed to improve their properties to suit other purposes. This can be done to make metals more aesthetically pleasing, to improve their strength or weight and to make them more corrosion- or scratch-resistant. Below are some more examples of commonly alloyed metals.

NAME	PROPERTIES	USES
Aluminium	Pure metal. Very good strength-to-weight ratio, and casts easily.	Drinks cans, window frames, food packaging, kitchen utensils.
Copper	Tough, ductile and malleable. Good conductor of electricity. Expensive.	Coins, plumbing/water pipes, electric wires and cables, jewellery.

NAME	PROPERTIES	USES
Brass	Non-ferrous. Corrosion-resistant, hard. Casts well; work hardens. Easily joined, polishes well, good conductor.	Hinges, door handles, screws, plumbing valves, taps, decorative items, boat fittings and castings.
Duralumin	Non-ferrous. Very good strength-to-weight ratio. Age hardens. Machines and finishes well.	Aircraft parts due to its very high strength-to-weight ratio.

THINGS TO DO AND THINK ABOUT

1. Describe the terms 'ferrous', 'non-ferrous' and 'alloy'.

2. Using the internet, find images of metal products, or find metal products around the home. Try to identify what types of metal have been used to manufacture them based on their colour, and think about why they were chosen. You should record this in your own revision notes.

DON'T FORGET

Learn the properties and uses for each category of metal, and be able to describe these when identifying what metals would be best for a given product.

VIDEO LINK

This video at www.brightredbooks.net looks at how steel is made and will give you an insight into the production of metals.

ONLINE TEST

Head to www.brightredbooks.net to test yourself on this topic.

PLASTIC

WHY CHOOSE PLASTIC?

Plastic is made by combining coal and crude oil, which are extracted from the earth and come in various forms such as: powder, granules, liquid, sheet and rod.

These forms make plastics highly versatile. Plastics are also easily moulded due to the advances in manufacturing technology, making them highly suited to mass manufacture. In many cases, plastics are a better choice of material and possess better properties than their wood and metal counterparts. With the continual development of new and improved plastics, they are the most preferred choice of material in product design. Designers may choose plastics due to any of the following general **advantages** that plastics offer:

- corrosion-resistance
- lightweight and strong
- recyclability
- easy to shape and mould; and thermoplastics can be reshaped/ remoulded
- variety of colours
- some plastics are available in translucent and transparent forms
- heat and electrical insulation.

This said, however, plastics also have **disadvantages** that we must consider:

- Some plastics can be difficult to cut, where they shatter or crack easily.
- Some plastics scratch easily, and these scratches are extremely difficult to remove.
- The production of plastics creates harmful by-products such as gases and chemical waste, which damage the environment.
- Plastics are not readily biodegradable and take hundreds if not thousands of years to break down. This results in pollution via landfills, which further damages the environment.

CATEGORIES OF PLASTIC

Plastic is divided into two main categories: thermoplastic and thermosetting plastics. Although there are other categories of plastic, you are only required to learn those two categories for your National 5 exam.

Thermoplastics

Thermoplastics are the most widely used plastic in modern manufacture. These plastics will soften when reheated and will return to their original shape. This is called **plastic memory**. Thermoplastics are often used in products that do not have to withstand high temperatures. There are many types of thermoplastic, all of which are suitable for different applications due to their unique properties.

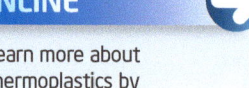

Plastic supply forms

ONLINE

Learn more about thermoplastics by exploring the link at www.brightredbooks.net

NAME	PROPERTIES	USES	NAME	PROPERTIES	USES
Acrylic	Stiff, hard, durable, scratches easily and brittle. Can be made opaque (solid colour) as well as translucent and transparent.	Good substitute for glass, lenses, signs, lighting, cases and jewellery.	**Low-density polythene (LDPE)**	Soft, flexible and a good electrical insulator.	Bags, plastic bottles, toy and cable covers.
ABS	Strong, light, durable, scratch-resistant and chemical-resistant.	Toys, kitchenware, plastic furniture, crash helmets and cases.	**Nylon**	Very hard-wearing, durable and rigid.	Gears, ropes, bearings and clothing/ upholstery.

contd

NAME	PROPERTIES	USES	NAME	PROPERTIES	USES
Polystyrene 	Light, stiff and water-resistant.	Food packaging, containers, games/DVD cases and protective packaging.	**Polypropylene**	Light, rigid and chemical-resistant.	Films, medical equipment, rope and kitchenware.
High-density polythene (HDPE)	Durable, tough and chemical-resistant.	Wheelie bins, buckets, bowls, sterilised containers and detergent bottles.			

DON'T FORGET

Learn the descriptions for each category of plastic, and be able to select plastics from these when identifying what plastic would be best for a given product.

ONLINE

Learn more about thermosetting plastics by exploring the links at www.brightredbooks.net

ONLINE TEST

Head to www.brightredbooks.net to test yourself on this topic.

Thermosetting

Unlike thermoplastics, thermosetting plastics will not return to their original shape or soften when reheated. Thermosetting plastics are often chosen for use in products that will be exposed to extreme temperatures, electrical currents, chemicals and sustained wear. Although thermosetting plastics can withstand high temperatures, they will not melt once overheated, and will fracture instead.

NAME	PROPERTIES	USES		NAME	PROPERTIES	USES
Melamine formaldehyde 	Waterproof; resists some chemicals; hard-wearing and scratch-resistant.	Laminated worktops, tableware (plastic cups, plates etc.), buttons and electrical insulation.		**Urea formaldehyde**	Good adhesive qualities, stiff, strong, hard, brittle and a good electrical insulator.	Electrical fittings such as sockets and plug casings. Handles, door knobs and adhesives.

THINGS TO DO AND THINK ABOUT

1. Describe the difference between a thermoplastic and thermosetting plastic.

2. Symbols are used to show what type of plastic has been used in a product and how it can be recycled. There are seven symbols. Using the internet, conduct a search on 'plastic symbols' to find out what each is. Record this in your notes, and then look at some plastic products found around the home and see if you can identify the plastic that has been used, looking for these symbols.

VIDEO LINK

This video on how plastics are made will give you an insight into the production and recycling of plastics. Watch the video clip at www.brightredbooks.net

MATERIAL PREPARATION AND FINISHING

WOOD

Wood naturally degrades, and therefore it must be protected to increase its lifespan, to improve its durability and to protect it from accidental user damage and environmental damage such as the weather when used outdoors.

Sandpaper

Sanding block

SANDING

To prepare wood for finishing, it must be sanded. This is done using sandpaper, also known as glass paper. Sandpaper is an abrasive paper for removing pencil marks and the majority of surface imperfections from wood. It comes in different grades ranging from coarse (rough) to fine (smooth). Coarse papers have a lower number ranging from 12 to 80, and fine papers have a higher number ranging from 100 to 220. To effectively sand wood, you must:

- Sand in the direction of the wood grain. This will provide the best finish, as going against the grain creates scratches that further damage the wood.
- Start by using a coarse-grade sandpaper to remove imperfections.
- Finish by using a fine-grade sandpaper.
- Sandpaper should always be wrapped around a sanding block (normally made from cork). This ensures the best finish, keeping the paper flat to the wood and stopping rounded edges from forming.

Obviously, the grade of sandpaper required will depend on the amount of imperfections to be removed.

FINISHES

Wood naturally degrades over time and will require finishing to improve durability. The environment and the product's function should be considered when selecting a suitable finish.

Varnish

Varnish

APPLICATION	PROPERTIES
Applied with a brush, brushing in the direction of the grain. Several coats may have to be applied, and the previous coat must be sanded with fine sandpaper before applying the next coat.	Water-resistant, glossy, enhances the grain and can alter the colour of the wood depending on the type of varnish used. Durable and hard-wearing, with infrequent application required.

Wax

Wax

APPLICATION	PROPERTIES
A sanding wood sealant should be used first. Once dry, the wax should be applied with a cloth, rubbing it into the wood.	Natural high-quality finish with a high sheen that enhances the natural colour of the wood. Will need frequent reapplication to maintain protection.

Oil

APPLICATION	PROPERTIES
Applied with a clean cloth, rubbing into the wood along the grain.	Enhances the natural finish of the wood. Reasonably durable and comes in differing depths of colour and types of finish. Requires reasonably frequent reapplication.

Oil

contd

Stain

APPLICATION	PROPERTIES
Applied using brushes or a cloth. Once the stain is applied, a suitable varnish, wax or lacquers can be applied to further protect the wood.	Comes in various colours that are used to colour the wood, making it look like a more expensive wood or just to add unnatural colour to the wood. Not very durable or protective, and requires further finishing.

Paint

APPLICATION	PROPERTIES
Paint is applied in the same way varnish is applied.	Colours the wood and often hides the natural grain. Best used on manufactured boards. Different finishes and colours can be achieved, and it provides a durable finish with infrequent reapplication required.

Lacquer

APPLICATION	PROPERTIES
Lacquers can be applied with a brush, but in commercial manufacture they are normally sprayed, giving the best possible finish.	Provides a glossy clear finish that enhances the natural qualities of the wood. Good durable protection, and hard-wearing. Guitar bodies are a good example of this finish

Unfumed

4 hours

8 hours

12 hours

16 hours

20 hours

24 hours

28 hours

32 hours

Wood stain

When applying any of the finishes listed above, it is important that brushes or cloths being used are clean. You should also take care to avoid over-applying a finish, and should be patient to let finishes fully dry before applying additional coats. When using brushes, remove any loose bristles that stick to the wood, and look out for runs caused by having too much on your brush.

Lacquered guitar

Paint finish

Veneers/Laminates

As mentioned earlier, manufactured boards are often unpleasing to the eye. For this reason, manufacturers often add veneers (thin sheets of softwood/hardwood) or plastic laminates to improve their aesthetics. Doing this can make a cheap MDF board look like expensive mahogany. Similarly chipboard, often used in desks/kitchen worktops, can be made to look like expensive marble where a plastic laminate is applied. Veneers/laminates are better for our environment: less sustainable materials can be recreated to give us the aesthetic qualities we want and at a much cheaper price.

Preparing to drill wood

Drilling into wood is normally easier than drilling into metal or plastic. Prior to drilling, the centre of the hole should be clearly measured and marked. To provide more accuracy, a **bradawl** can be used to pierce the centre of the hole before drilling. When drilling a bigger hole, it is good practice to pre-drill a smaller hole called a **pilot hole**. This will stop the wood from splitting and will make it easier for the drill to cut the bigger hole. Pilot holes can also be drilled in plastic and metal.

Bradawl

DON'T FORGET

You must be able to explain the sanding process and know the suitability and application of each wood finish.

ONLINE

Learn more about wood finishes at www.brightredbooks.net

VIDEO LINK

Check out the clip at www.brightredbooks.net for more on this topic.

THINGS TO DO AND THINK ABOUT

1. Describe the key stages that must be observed when sanding.

2. The specification requires a child's seat that is made from wood and is hard-wearing, colourful and durable. Select a suitable material and finish that would help the designer to achieve this.

ONLINE TEST

Test yourself on wood finishing at www.brightredbooks.net

METAL AND PLASTIC

RECTANGULAR

profile

HALF-ROUND

profile

ROUND

profile

SQUARE

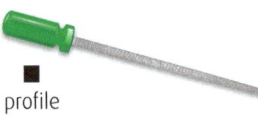

profile

TRIANGULAR

profile

METAL

Most metals need to be cleaned before finishing, because certain metal processes often require grease or oiling when working with the metal. Metal should be well cleaned before applying any finish.

Filing

Metal often has rough or sharp edges which will need to be filed to smoothen them. Filing is carried out using a file. There are several different types of file, including those shown left:

The process of filing when finishing is carried out as follows:

1. **CROSS-FILING:** Using a coarse file, file across the edge of the metal, removing all scratches and deformations (i.e. deep dents).

2. **DRAW FILING:** Using a smooth file, file lengthways along the edge of the metal to smoothen out the marks left from cross-filing.

3. **EMERY CLOTH:** Wrap a piece of emery cloth (abrasive paper used for metal) around a file, and follow the procedure for draw filing to further smoothen the metal.

4. **STEEL WOOL:** Rub steel wool along the edge of the metal to fully smoothen.

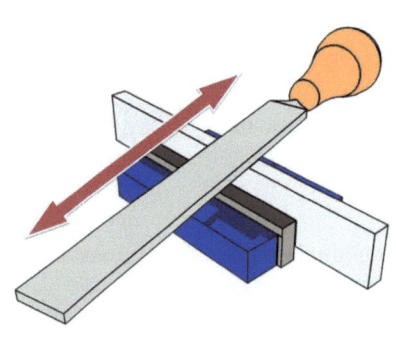

Cross-filing

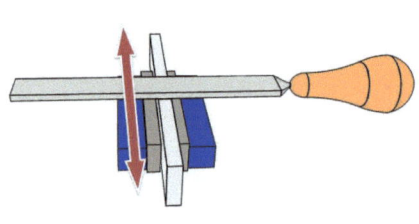

Draw filing

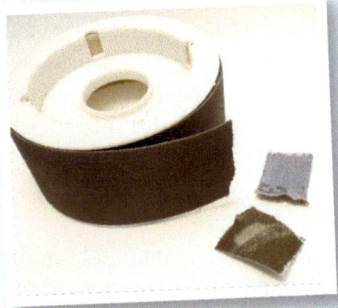

Emery cloth

Steel wool

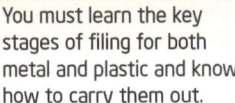

DON'T FORGET

You must learn the key stages of filing for both metal and plastic and know how to carry them out.

Preparing metal for drilling

The difficulty of drilling varies for each type of metal. The drill should be set to the correct speed, and the correct type of bit must be used. As metal has a hard, smooth surface, drill bits often slip when they make contact with the material. It is important that the centre is marked with a centre punch before drilling. This creates a small dent in the centre position of the hole, allowing the drill to find the centre easily.

A lot of friction can be generated when drilling metal, which can burn out drill bits. Therefore it is important that you do not force the drill bit through the material. Grease or a cutting agent can also be applied to make the process easier and reduce the friction.

contd

Finishes

Ferrous metals in particular need protection, as they rust due to their iron content. Non-ferrous metals can often be left unfinished due to their natural finish; and alloys require little finishing, as they are often designed to look that way.

FINISH	APPLICATION	PROPERTIES
Painting	Paint is applied with a brush. The finish shown here is a hammerite finish.	Colours the metal and protects it from rusting. Different finishes and colours can be achieved, and it provides a durable finish with infrequent reapplication required.
Lacquer	Lacquers can be applied with a brush, but in commercial manufacture they are normally sprayed, giving the best possible finish.	Provides a glossy clear finish that enhances the natural qualities of the metal. Good durable protection, waterproof, protects from rust and is hard-wearing.
Plastic dip coating	Plastic powder is placed in a **fluidiser**. Cold air is blown in, which causes the plastic to begin bubbling, making it look like it is boiling; however, it is not and is cold. The metal is heated in an oven, and, once at the correct temperature, it is dipped into the plastic powder. The air blown into the fluidiser allows the plastic to coat the metal evenly and begins to cool it. *A video is provided to improve your understanding of this process and the fluidiser machine used.*	Once set, a hard-wearing layer of plastic with a good sheen is coated over the metal. This is useful for handles.

VIDEO LINK

Check out the clip on plastic dip coating at www.brightredbooks.net

ONLINE

Follow the link on the Digital Zone for more on this.

PLASTIC

Due to the way plastics are manufactured, they require very little finishing, being durable, waterproof and available in a variety of colours. However, there are still some techniques that must be applied to finish plastics. The **edges of plastics** should be filed as outlined previously for metal finishing, using **wet and dry paper** instead of emery cloth. **Wet and dry paper** is an abrasive paper used for plastic. It works best when used wet, as this prevents the clogging of plastic fibres, producing a smoother finish. To fully finish plastic surfaces after working, a suitable **plastic polish** can be used. This is rubbed into the plastic using a cloth, and will help to remove some minor scratches, ensuring the surface has a high shine.

Preparing plastic for drilling

Plastic can be difficult to drill, as certain plastics can crack easily. As with wood and metal, the centre of the hole should be clearly marked. It is important that you go slowly when drilling plastic to avoid cracking. Masking tape can also be used to cover the area around the hole, offering some protection. A piece of wood should be placed underneath the plastic to reduce the impact when the drill comes out the other side.

DON'T FORGET

You must learn all of the finishes listed, how they are applied and how they protect the material.

ONLINE TEST

Test yourself on metal and plastic finishing at www.brightredbooks.net

THINGS TO DO AND THINK ABOUT

A video is provided in the Digital Zone for plastic dip coating. Using YouTube, search for the other finishes and preparation techniques described on this page to help you improve your understanding of them.

PLANNING FOR MANUFACTURE

STAGES OF PLANNING

Before manufacturing a product, it is extremely important that you are well prepared. This will limit mistakes during the manufacturing process, reducing waste and ensuring the project is completed to the correct standard, and on time.

Selecting materials

This stage involves selecting the best materials for the job. Once the designer has decided on which materials to use, the next stage is to source them. Working with an engineer, a cutting list will be produced such as the one given here used to design the wooden box shown.

Part	Name	Material	Length	Width	Thickness	No. required
A	SIDES	PINE	250	100	12	2
B	FRONT/BACK	PINE	150	100	12	2
C	BASE	PLYWOOD	238	138	4	1
D	LID	PINE	250	150	12	1

Wooden box

A cutting list outlines the number, size and type of materials required for each part. The box shown will also require standard components such as the locking mechanism and hinges. See pp. 38–39 for more information on selecting materials and standard components.

Preparing the material

Once the materials are purchased, they may need to be prepared. This varies for differing materials and includes cutting materials to size, smoothing surfaces, removing saw marks, cutting the material to required shapes/patterns using templates and removing any imperfections where possible.

Planning: sequence of operations

This may be done before the materials are prepared, but you must already know what materials you are using before you can accurately decide on suitable manufacturing processes and techniques. Planning a sequence of operations requires manufacturers, in consultation with designers and engineers, to list all of the steps required to manufacture the product from start to finish. It is important to establish the most efficient and cost-effective way to build the design. Planning includes:

- how the materials will be prepared
- what tools, moulds, jigs or templates will be required for the chosen processes
- how the manufacturing processes will be carried out
- how the materials will be assembled
- what finishes will be applied.

MANUFACTURING ACTIVITES FOR THE WOODEN BOX

1. Ensure all wood is of the correct size and type and is fully prepared for manufacture.
2. Mark out corner rebate joints on both sides using a pencil, try square, marking gauge and steel rule.
3. For each corner rebate joint, ensure the correct width for the adjoining wood is measured and that the depth of the joint is marked no more or less than halfway.
4. Cut the width and depth of the corner rebate joint using a tenon saw to remove waste wood. A jig may be required to ensure lines are cut straight and accurately.
5. Use a bevel-edged chisel to level the joint where required.
6. Use a rebate plane, set to the correct depth and width of the plywood base, to cut a rebate along the bottom edges of both the sides and the front and back.
7. Assemble the box and dry-clamp, checking if the box is square and of the correct assembly size.
8. Prepare wood for assembly, sanding all surfaces. All marks/imperfections should be removed.
9. Use PVA glue in all joints and clamp together using sash cramps. Ensure the box is square.
10. Attach the base to the box using PVA glue and panel pins. Panel pins should be secured using a pin hammer.
11. Mark out position for hinges and lock mechanism on the box and lid.
12. Drill pilot holes where required for the screws that secure hinges and lock mechanism.
13. Attach the hinges with the correct type of screwdriver.
14. Final preparation for chosen finish through light sanding.
15. Apply finish as required. Additional coats should be applied correctly through sanding and finishing until desired finish is achieved.

Working drawings and diagrams

To help manufacturers understand how the product will be made, engineers will create production engineering drawings, also known as working drawings. One of the key types of graphic produced at this stage is an **orthographic drawing**. Two main types of orthographic drawing are often drawn, namely **component** and **assembly** drawings. **Component orthographics** are produced to show specific details for each individual component of a product. These will include detailed sizes and manufacturing notes. **Assembly orthographics** are used to show how the product is assembled and will not contain many, if any, dimensions.

Orthographic drawing is a graphics technique where views of a 3D object are drawn in 2D, looking straight on at every side of the object. The most common views are

contd

the **elevation** (front), **end elevation(s)** (side(s)) and the **plan** (top). The elevation is the first view to be drawn, with the end elevations positioned directly to the side of this and the plan view directly above it. This layout method is known as **third-angle projection**. Orthographic drawings are extremely useful in providing a working drawing, as engineers can easily show key features of the product and include precise dimensions (sizes).

To further explain the working of the drawing, engineers will also use pictorial views, exploded views and sectional views as explained on pp. 16–17 and will also include assembly/production notes detailing what has to be done.

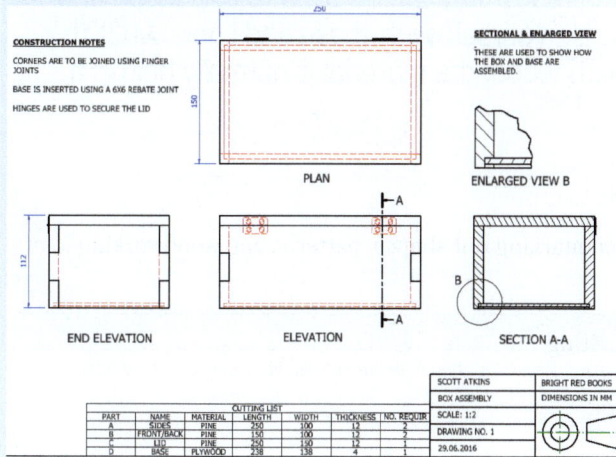

Orthographic drawing of the wooden box

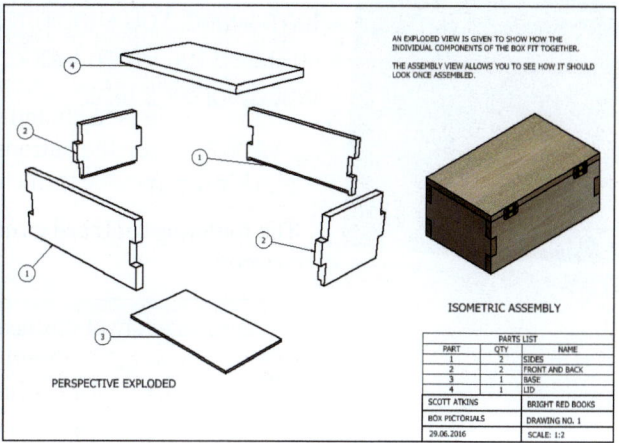

Pictorial views of the wooden box

SAFE WORKING PRACTICES

When undertaking any manufacturing activities, it is extremely important that safety procedures are followed accordingly. Failure to follow safety practices can result in injury for those involved in the manufacturing process. Therefore there are standard safety protocols that must be followed:

- Check all tooling, machinery and equipment is in good working condition before using it.
- Maintain all tooling, machinery and equipment
- Wear correct safety equipment as is appropriate for the manufacturing activity.
- Be extra-vigilant when working, i.e. acting responsibly at all times, looking out for the safety of others, taking off jewellery, tying long hair back/securing loose clothing.
- Report any accidents, damage or breakages immediately.

It is important that these procedures are followed and built into the planning-for-manufacture process to ensure safe working practices at all times.

To ensure safety, there are a number of clothing items we can use, and signs that alert us to safety issues:

Safety Clothing		Safety Signage	
Apron	Protects clothing, e.g. fire resistant aprons used for working with the forge and welding equipment.	**Blue**	Mandatory: you must do something, i.e. what safety equipment to wear.
Gloves	Protects hands when working with metals or using guillotine, forge and welding equipment.	**Yellow**	Warning of danger, i.e. of electrocution.
Safety Glasses	Standard protection for eyes	**Red**	Prohibition signs, warning you not to do something.
Face shield	Protects entire face and mainly used when working on the lathe.	**Green**	Safety signs such as fire exits and first aid.
Welding mask	Protects eyes from exposure to extreme light produced when welding.		

THINGS TO DO AND THINK ABOUT

Using the manufacturing activities diagram on p50 as reference, write out a sequence of operations for a product you are going to make in school. Similarly, write one for a product found at home or that you have made before. Consider the sequence of operations produced prior to manufacture.

COMMON WOODWORKING TOOLS

At National 5 level, you must know the tools used to work wood, the methods of constructing/joining wood and the process of wood-turning.

There are various tools and processes that can be used to cut, shape and form wood. You should familiarise yourself with these and know what they are used for. When working with wood, the material is normally held in a woodworker's vice.

Marking out

The following tools are used when marking out shapes, patterns and woodworking joints in wood.

TOOL NAME		PURPOSE
Steel rule		Used for measuring. Steel rules often measure in millimetres (mm) for more accurate measurement.
Try square		Used for marking straight lines at 90° to the edge of a piece of wood. When using the try square, you must hold the handle flat against the edge of the wood to ensure accuracy.
Sliding bevel		Used for marking and measuring various angles.
Marking gauge		Used to mark straight lines parallel to the edge of a piece of wood. When using the marking gauge, it is important the stock is flat against the edge of the wood. When measuring, you should always measure against the point of the spur, as this is the part that will mark the wood.
Mortise gauge		Used for marking the parallel lines required for accurate positioning of a mortise and tenon joint. The mortise gauge has two pins, one of which is fixed and one that is adjustable. This allows the manufacturer to set the correct size required for the mortise of a mortise and tenon joint.

CUTTING AND SHAPING TOOLS

Saws

Sawing is a common process for cutting and shaping wood. There are two categories of saws: **rip saws** and **cross-cut saws**. Rip saws are used for cutting along the grain, which is an easier job requiring less effort, whereas cross-cut saws are used for cutting across the grain, which is the more difficult task. Cross-cut saw blades have a sharp-angled edge on the teeth to help make cutting across the grain less labour-intensive. Many of the saws below are available in both formats.

When sawing, saws remove a section of material the exact width of the saw teeth. This is referred to as the saw **kerf** and tells us how much of the material will be lost due to sawing.

TOOL NAME		PURPOSE
Panel saw		Used for cutting larger pieces of timber to size.
Tenon saw		Carpenter's saw used for accurately cutting joints. The tenon saw offers more accuracy due to the hard brass back that runs along the length of the blade. This helps to ensure that the saw will remain straight when cutting.
Scroll/ fret saw		A scroll/fret saw can be described as an electric version of a coping saw. The blade can be set at various speeds, providing quick and accurate cutting for wood and plastics. It can also be used to easily cut curved/awkward profiles and is also useful for cutting materials such as foam for modelling.
Mitre saw		Used for cutting mitre joints at 45°, although the angle can be changed for different cuts.
Coping saw		Used for cutting shapes in wood such as curves or sections that need to be removed from the middle of a piece of wood. The coping saw is small and the blade is delicate. Coping saws can also be used for cutting plastics.
Bench hook		A bench hook is used to provide a stop, where a piece of wood being worked can be placed against it. This allows you to steady the piece of wood when sawing.

contd

Chisels

There are a variety of chisels that can be used for different purposes. Chisels are generally used to remove waste wood from woodworking joints. They can be controlled by hand; and sometimes a mallet may be required for extra force when working with tougher materials or cutting out more complex joints or patterns.

TOOL NAME	PURPOSE
Firmer chisel	Used for general chiselling work to remove waste material from woodworking joints.
Bevel-edged chisel	Used for general chiselling work; however, the bevelled edges on this chisel make it particularly effective when trying to cut into tight corners.
Mortise chisel	Mortise chisels are specifically used for chiselling out the mortise for a mortise and tenon joint. This joint requires a lot of force to cut out. A mallet is often used along with a mortise chisel, and for this reason mortise chisels have a steel ferrule at the top of their handle to stop it from splitting.
Beech mallet	Used when extra force is required to chisel out waste material.

Planes

Planes are used for shaping, finishing, smoothing, reducing the wood's thickness or width, curving or chamfering edges and cutting some joints. Each plane's blade can be adjusted for different depths of cut.

TOOL NAME	PURPOSE
Jack plane	Used for smoothing down surfaces and reducing the thickness of larger pieces of timber.
Smoothing plane	Smaller version of a jack plane used for smaller jobs.
Rebate plane	Used for producing shoulders or rebates.
Plough plane	Used to cut grooves into the wood. These grooves may be used for inserting hardboard drawer bottoms or bases in carcase constructions.
Router plane	Normally used for finishing work such as clearing out the bottom of housing joints.

THINGS TO DO AND THINK ABOUT

1. To help prepare for your exam, you should begin producing rote answers that describe how to carry out cutting and shaping processes in wood using the tools above. This could involve answers on marking and cutting different woodworking joints, cutting curves or awkward shapes in wood, and planing wood. See pp. 56–57 on wood joints, where an example question is given for manufacturing a wood joint, to help you prepare these answers.

2. Using YouTube, you should search for videos related to the above tools. This will help you to develop a better understanding of how they are used.

DON'T FORGET

You may not have to explain these tools in the exam, but you should learn what each tool's purpose is and be able to refer to the correct tool when explaining how to carry out processes in woodworking.

VIDEO LINK

Check out the clips at www.brightredbooks.net for advice on using woodworking tools.

ONLINE TEST

Test your knowledge of common woodworking tools at www.brightredbooks.net

WOODWORKING JOINTS: CARCASE CONSTRUCTION

Disassembled butt joint

Disassembled mitre joint

Assembled mitre joint

Disassembled corner rebate joint

Assembled corner rebate joint

INTRODUCTION

There are several ways in which we design wood to be joined. Woodworking joints are a traditional method of joining wood, and there are many different types that serve a variety of purposes. Joints are often cut by skilled tradesmen, but modern technology now affords the opportunity to use automatic CNC (computer numerical control) machines. This ultimately speeds up the process and can increase accuracy, giving no risk of human error.

There are two main categories of woodworking joints used in construction. These are known as **carcase** and **frame** joints.

CARCASE CONSTRUCTION

Carcase construction refers to any wooden product that is constructed in a box style. This could include furniture such as wardrobes and cupboards. The following carcase joints can be used in carcase construction.

Butt joint

This is a very weak joint where the wood is simply butted against another piece and glued. It can be strengthened using pins or screws.

Mitre joint

Used to join corners only, a mitre joint is created by cutting both adjoining pieces at 45° to each other and then butting and gluing them together. It can be further strengthened using pins, heavy-duty staples or screws.

Corner rebate joint/lap joint/corner housing joint

Used to join the corners of carcases together, the corner rebate joint is reasonably strong. A channel is cut along the end of one piece of the wood being joined to no more **or** less than half the depth of the wood, and to the full width of the adjoining piece of wood. This ensures structural strength.

Housing joints

Used to join partitions or shelves together, housing joints are very strong. There are two types of housing joint: **through** and **stopped**. Both offer the same structural properties; however, stopped housing joints are often the preferred choice as they hide the workings of the joint, making them more aesthetically pleasing. Housing joints are constructed by cutting a channel along one piece of wood to no more **or** less than half the depth and to the full width of the adjoining piece of wood, to ensure structural strength. In a stopped housing joint, the channel is not cut all the way through, creating a stop to hide the working of the joint.

Disassembled through housing joint

Disassembled stopped housing joint

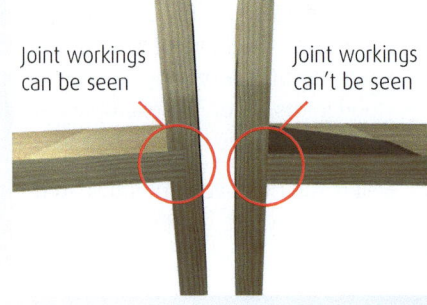

Joint workings can be seen

Joint workings can't be seen

Assembled through vs stopped housing joints

contd

Dowel joint

Dowel joints are made using small wooden rods known as dowels. A hole is drilled in each piece of the wood to be joined, to a depth representative of the size of dowel being used and thickness of the wood. The wooden dowel is then glued into both holes and the wood joined together. The width of the wood will determine the number of dowels required. As a rule, two is the minimum, as this will ensure the joint is strong enough and will not turn or move while gluing. To ensure that dowelling holes line up, dowel marker pins can be used for accuracy. These are placed in the first hole that is drilled, and then the adjoining piece of wood is pressed into these to mark the positions of the holes for drilling.

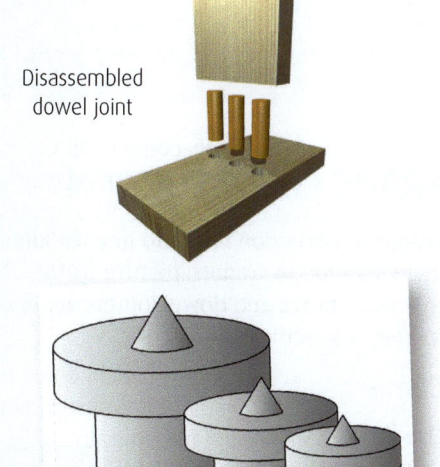

Disassembled dowel joint

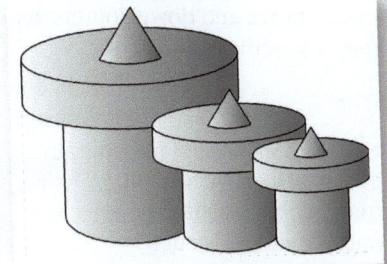

Dowel marker pins

Comb (finger) joint

A comb joint, also known as a finger joint, is a very secure way of joining corners in carcase construction. Adjacent square or rectangular sections are removed from each piece of wood to create an interlocking system, much like if you were to lock your fingers together when joining your hands. Comb joints offer good strength and can be used as an aesthetic design feature.

Disassembled comb joint

Assembled comb joint

Dovetail joint

The dovetail joint is similar to a finger joint but offers much more strength. Rather than cutting out adjacent square or rectangular sections, dovetail joints are cut out using angled sections. This offers high structural strength. Dovetail joints can also be used to add aesthetic value to a design.

Disassembled dovetail joint

Assembled dovetail joint

VIDEO LINK

To see how to make a mitre, a corner rebate (lap) and a through housing joint, watch the video clip at www.brightredbooks.net

DON'T FORGET

In the exam, you may be asked to identify a joint for manufacture of a given product. You may also be asked to describe how that joint would be manufactured.

THINGS TO DO AND THINK ABOUT

For each of the carcase joints listed here, prepare answers on how to manufacture them, including the tools that would be required to do so. This will help you in preparing for answering questions in the exam that ask you to describe how a joint would be manufactured. See pp. 56–57 on wood joints, where an example question is given for manufacturing a wood joint, to help you prepare these answers.

ONLINE TEST

Head to www.brightredbooks.net to test yourself on this topic.

WOODWORKING JOINTS: FRAME CONSTRUCTION

FRAME CONSTRUCTION

Frame construction refers to any wooden product that is constructed as a frame. This includes window frames, picture frames, table and chair frames. It is important to note that **butt**, **mitre** and **dowel** joints can all be used in frame construction as well as the following specific frame joints:

Corner halving joint

Used to join the corners of frames together, the corner halving joint is reasonably strong. A channel is cut along the end of one piece of the wood being joined to no more **or** less than half the depth of the wood, and to the full width of the adjoining piece of wood. This ensures structural strength.

Disassembled corner halving joint

Assembled corner halving joint

Tee halving joint

Used to join internal parts of the frame, e.g. the middle sections of a traditional window frame. It is manufactured in the same way a corner halving joint is, with the only difference being the position.

Disassembled tee halving joint

Assembled tee halving joint

Cross halving joint

Used to join parts of the frame that must cross one another. These are manufactured in the same way that corner and tee halving joints are; however, both pieces of wood are cut in the exact same way to allow them to cross and join.

Disassembled cross halving joint

Assembled cross halving joint

Mortise and tenon joint

Perhaps the strongest frame joint, the mortise and tenon joint can be used for corners and partitions. A mortise (hole) is cut in one piece of wood manually using a mortise chisel and mallet, or automatically using a mortise machine. The tenon (connecting piece) is cut out using a tenon saw to create a section of wood on the adjoining piece that fits the exact size of the mortise.

Disassembled mortise and tenon joint

Assembled mortise and tenon joint

DON'T FORGET

You must learn the properties and purpose of each joint, and be able to explain how they would be manufactured manually.

ANSWERING QUESTIONS ON WOODWORKING JOINTS

In the exam, you are likely to be asked questions on how particular joints are manufactured. The question will normally refer to a specific product and indicate what joints have been used. You will have to be able to describe how the joint is cut. To answer these questions effectively, ensure that:

- you know each of the joints described above
- you know the common woodworking tools required to cut the joint
- you have prepared answers that describe how each joint can be manufactured.

An example is given below to help you in preparing for these types of question.

EXAMPLE:

1. A toy truck manufactured from wood is shown here.
(a) Name a suitable corner joint for the construction of the back trailer. (1 mark)

Advice: You could select a butt joint; however, this would not be durable enough to survive the kind of use it would be subjected to by a child. You could also select a finger joint, which would offer good durability; however, this could increase the cost of manufacture, and the aesthetic value offered would not be of high importance to a child.

ANSWER:

Lap joint. This would be suitable, as it is relatively cheap and quick to produce, is reasonably strong and can be further strengthened by pins or screws to increase its durability.

(b) Describe how the joint you selected previously would be manufactured. (3 marks)

Advice: if the question is worth three marks, then try to make three or more points that explain how the joint would be manufactured. Remember that every joint will require:

- marking out
- cutting
- finishing and gluing.

You can use these four points to build your answer, and refer to tools as you do so.

ANSWER:

1. *Mark out the area to be removed on the end of the required piece of wood using a rule, try square and marking gauge. Ensure the markings are: no more than half the depth, the correct width of the wood to be joined, square and parallel to the edges.*
2. *Using a tenon saw, cut out the area to be removed by firstly sawing down no more than halfway. Then saw down the line used to mark the length of the joint, to remove the waste material.*
3. *Using a bevel-edged chisel, remove any excess material to ensure that the joint surface is flat.*
4. *Finally, assemble, glue and clamp the joint.*

Corner joint

ONLINE

Learn more about wood joints by exploring the link at www.brightredbooks.net

ONLINE TEST

Head to www. brightredbooks.net to test yourself on this topic.

THINGS TO DO AND THINK ABOUT

1. For each of the frame joints listed here, prepare answers on how to manufacture them, including the tools that would be required to do so. This will help you in preparing for answering questions that ask you to describe how a joint would be manufactured.
2. Using YouTube, search for the names of some of the joints listed over the previous few pages. You will find videos that will explain how these joints are cut and assembled.

WOOD PROCESSES: TURNING

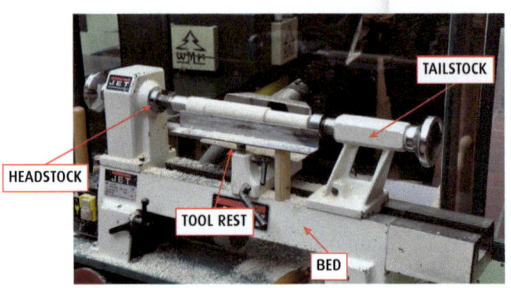

Standard wood lathe

INTRODUCTION

A wood lathe is used to turn lengths of wood into cylindrical forms, for example candlesticks, tool handles, or spindles for handrails. It can also be used to make bowls using face plates. Lathe work can be carried out manually or as a fully automated process using CNC lathes controlled by computers. The process can vary in cost according to labour and machine parts. If turning is carried out manually, it is only suited to low-volume production, whereas if it is automated it can be used for much higher volumes. When turning, it is important that the speed of lathe is correct for the size and type of material being turned, as well as the process being undertaken. This ensures a good-quality finish and limits any potential risks.

SETTING UP THE LATHE

When setting up a wood lathe, the following must be done:

- Wood prepared for turning (explained later in this chapter).
- Ensure you have marked out 'dead areas' at the ends of the wooden blank. These are areas that will not be worked (cut) and will ensure safety in stopping the chisels from coming into contact with the fork or centre.
- Secure the prepared wood centrally to the **fork** and **centre**. The fork has four prongs that grip the wood and a centre point to ensure the wood is centred. The fork is held in the headstock. Centres are held in the tail stock and two types can be used. A **revolving centre** spins, reducing the friction on the wood meaning that turning can be carried out at higher speeds. A **dead centre** does not spin and creates significant friction. A lubricant should be used when using a dead centre.
- Set the vertical height of the tool rest to the correct height for the type of chisel and process being used. This is often just below the centre of the wood's width.
- Set the position of the tool rest in between the centres of the lathe and ensure it is kept at a safe distance away from the wood.
- Ensure the lathe speed is correct for the type of wood being used and the process being carried out.
- Ensure you have the correct safety equipment on i.e. apron and face shield, and that you adhere to all safety requirements such as tying long hair/loose clothing back.

1. Gouge:
Used for removing waste and rough cutting (roughing). Spindle gouges can also be used for creating coves and rounded grooves.

2. Parting:
Used for cutting notches or grooves and for parting (separating) the wood.

3. Scraper (round):
Used for removing waste wood, but provides a better finish than the gouge. It can also be used to create convex curves.

4. Skew:
Used for accurate cutting and shaping and for cutting 'V' grooves or design features into the wood.

TURNING TOOLS

There are a number of turning **chisels** that can be used for turning wood. When turning, these chisels are held using both hands and securely rested on the tool rest. The user can then move the chisel into and along the wood in different directions and at different angles to make cuts as required, shaping the wood.

To secure the wood in the lathe, the following tools are required:

Fork

The fork has four prongs that grip the wood and a centre point to ensure that the wood is centred. The fork is held in the headstock.

Revolving centre or dead centre

Both centres are used to hold the wood centrally. However, a revolving centre spins, reducing the friction on the wood, meaning that turning can be carried out at higher speeds. A dead centre does not spin and therefore creates significant friction. A lubricant should be used when using a dead centre.

Lathe set up

PREPARING WOOD FOR TURNING

Before turning a piece of wood, it must be prepared. To do this, the following steps must be taken:

1. First prepare the piece of wood, also known as the blank, by marking the diagonals from corner to corner on both ends of the blank. This makes it easy to locate the exact centre.
2. Saw down no more than 5 mm on both lines at one end of the blank, using a tenon saw to create two deep grooves.
3. Using a marking gauge, mark parallel lines down the length of the blank. Use a jack or smoothing plane to plane the corners down to the lines, or cut off using a bandsaw. This will reduce the resistance when you begin turning.
4. Locate the two grooves into the fork, and locate the centre mark at the other end of the blank with the tip of the revolving or dead centre.

 Step 1
 Step 2

 Step 3
 Step 4

WOOD LATHE PROCESSES

Parallel turning

Parallel turning is often carried out with a skew chisel for accuracy and quality of finish. To parallel turn, the chisel (rested on the tool rest) is kept at a steady and constant distance from the wood when taking the cut. The chisel is then moved from left to right and right to left at a steady/constant speed taking a parallel cut from the wood.

Parting off

Parting off is a process used to: square off the end, or features of wood turning, cut square edges grooves (trenches) and separate the finished job from the wooden blank dead areas. This process is carried out with a parting off chisel. The chisel is held at a right angle to the wood and the cut taken by feeding the chisel directly into the wood.

Finishing

Sanding: Due to the complexities of the turned piece of wood, finishing by sanding can be difficult. Therefore, sanding is often carried out on the lathe. To do this, the tool rest is removed from the machine and a piece of sand paper, held firmly in the hand, is pressed against the wood whilst it is turning. The sand paper is then moved left to right/right to left, creating a smooth finish.

Waxing: To complete the finish turned pieces of wood are often waxed using a hardened block of wax. The wax is pressed into the wood whilst the lathe is turning and moved left to right/right to left creating an even finish. Wax can also be applied with a cloth using the same method for sanding.

Materials used

Turning can be carried out using a variety of solid woods. It is important that the wood being used does not have any cracks or previous damage, as this could cause the wood to split during turning.

Products made and identification

Products made on a wood lathe include handrail spindles, tool handles, candlesticks and bowls. Products turned on a wood lathe will be cylindrical.

THINGS TO DO AND THINK ABOUT

Explain the process of preparing a wooden blank before turning.

DON'T FORGET

Ensure you can describe each of the areas outlined on this page as you may be asked to describe these in the exam.

VIDEO LINK

Check out the clips at www.brightredbooks.net for more on bowl-turning, parallel turning, parting off, sanding and finishing.

ONLINE

Check out the links at www.brightredbooks.net for more on basic turning tools and using a wood lathe.

VIDEO LINK

The videos at www.brightredbooks.net provide good examples of how wood is turned on a lathe and of automated woodturning.

ONLINE TEST

Test yourself on this topic at www.brightredbooks.net

FORMING AND JOINING WOOD 1

Solid woods are often joined using the wood joints outlined on pp. 54–57. However, to fix these, we need clamps and glue. The glue we use when joining wood is **PVA glue**. Screws, nails or pins may also be used to strengthen these joints. Wood joints, however, are often not suitable when working with manufactured board, and therefore screws, pins and nails will be required as well as knock-down fittings.

CLAMPING

Before gluing woodwork joints, we should always **dry clamp** first. Dry clamping means assembling without glue. This is useful for checking that the pieces being fitted together fit correctly and for testing how many clamps are needed to assemble the project. To do this, the following clamps can be used:

Sash cramps in use

Sash cramp: long enough to allow various sizes of projects to be glued.

G-clamp: can be used for gluing small projects together – but remember they are also used for securing wood to a workbench when working it.

Fast clamp: for quick clamping of small projects that do not require a lot of force to secure them. They can also be used for securing wood when working it.

Problems can occur when clamping, and **gluing checks** must be carried out to ensure the assembly is correct.

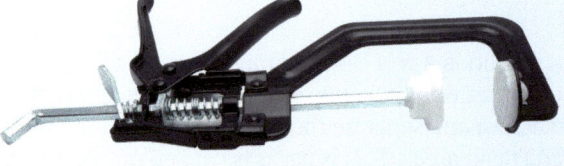

GLUING CHECKS

1. **Square:** A try square should be used to check all corners are at 90°. You must also check the diagonals measuring from one corner to the other corner directly opposite. If both diagonals are the same, then the wood is square.

If the job is not square, then **to fix this,** you may need to readjust the clamps, loosening or tightening them to pull the wood square. This could also be caused by joints not being of the same depth or size.

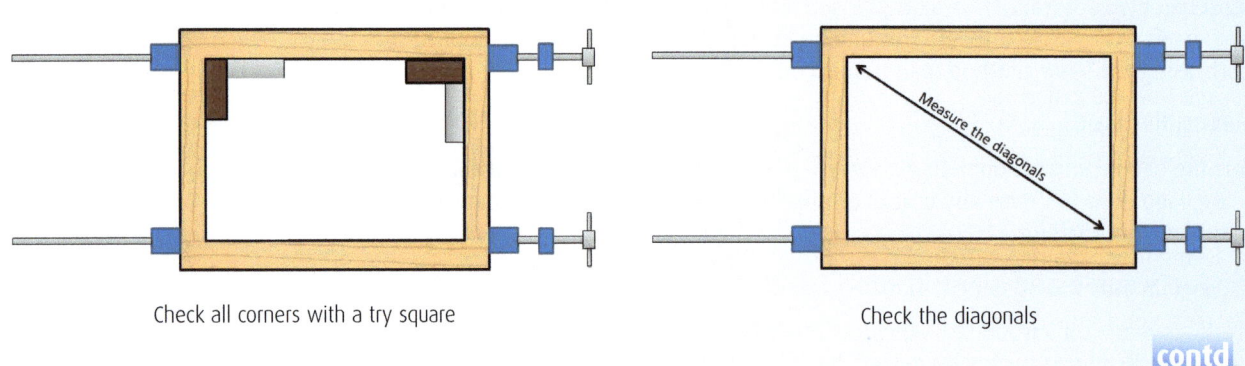

Check all corners with a try square

Check the diagonals

contd

2. **Cupping:** This happens when the clamps are over-tightened, causing the wood to lift due to the force being applied at either side.

To fix this, you should loosen the clamps until the wood lies flat, and add weights to the top to weigh it down.

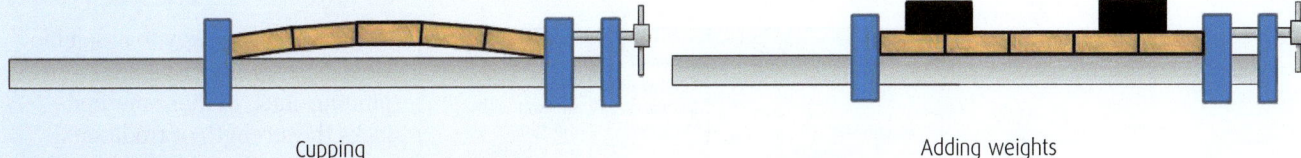

Cupping   Adding weights

3. **Twisting:** If the wood is not held flat, or the clamps are not level with each other when clamping, the job can twist.

To fix this, loosen the clamps and hold the wood flat before retightening.

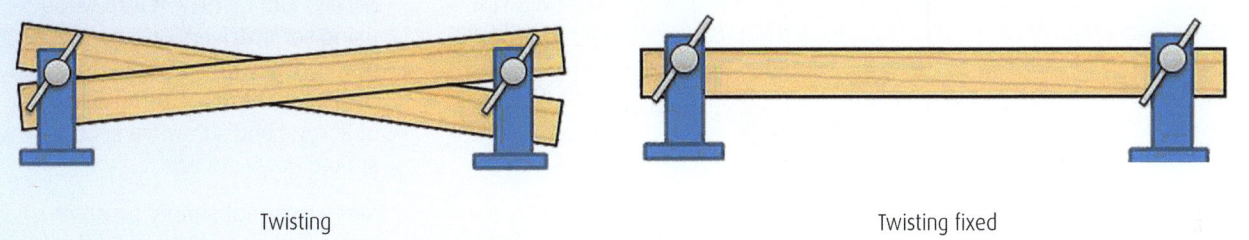

Twisting   Twisting fixed

Finally, you should make sure that all parts are assembled correctly and aligned as planned. Once dry clamping is correct and you know how to clamp the job, you can glue the project. Once glued, make sure any excess is wiped off.

LAMINATING

As solid timbers do not come in board sizes, we often have to glue them together when making larger projects. To do this, the process of laminating is used, which involves gluing the wood side by side. When doing this, it is extremely important that the end grain is glued in opposite directions to each other. This ensures that the wood doesn't cup as described above, and provides the best strength properties.

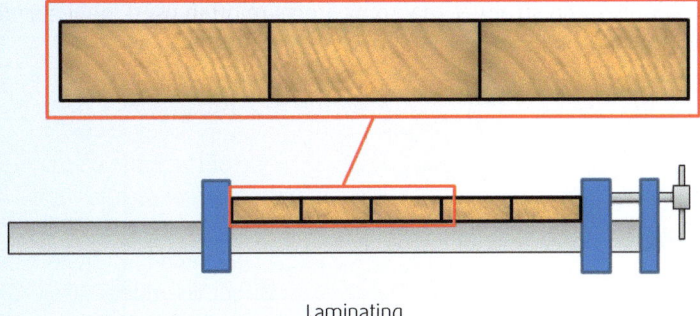

Laminating

THINGS TO DO AND THINK ABOUT

1. Describe what gluing checks you would carry out to ensure a box was square before gluing it together.

2. When in class, see if you can practise some of the techniques for gluing when dry-clamping your projects.

ONLINE

For more information about clamping check the link at www.brightredbooks.net

ONLINE TEST

Test yourself on this topic at www.brightredbooks.net

DON'T FORGET

Make sure you can explain the purpose of 'dry-clamping' and understand the process of checking if wood is square, level and true.

FORMING AND JOINING WOOD 2

SCREWS, NAILS AND PINS

	NAME	PURPOSE
	Countersunk screws	These screws sit level with the surface of the wood in a countersunk hole.
	Roundhead screws	These screws sit above the surface and can provide an aesthetic industrial quality.
	Philips screwdriver	Used to screw in screws with cross heads.
	Slotted screwdriver	Used to screw in screws with straight heads.

A quick and easy way to assemble wood is screwing or nailing/pinning. Each of these methods lacks the strength of traditional wood joints. In combination with wood joints, however, these can really strengthen an assembly.

Screwing

Screws can be driven into wood using screwdrivers or a cordless power drill with specially designed drill bits. There are different types of screws and screwdrivers that you should be aware of:

Screws cannot simply be driven into wood, as the pressure they cause when tearing into the wood will cause it to split. Therefore, the following drilling procedures must be carried out:

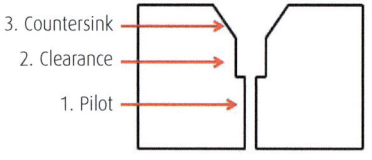

1. Drill a **pilot hole** (2–3 mm), which will reduce the stress on the wood when the screw goes in and will create a guide.

2. Drill a **clearance hole,** which allows the screw to fit the hole before being screwed in.

3. **Countersink** the top of the clearance hole if using countersunk screws.

Nailing/pinning:

Nails and pins are both driven into wood using a pin hammer (Warrington hammer). Nails are often longer and stronger than pins. Pins are often used for securing bases and backs for drawers, boxes and cupboards.

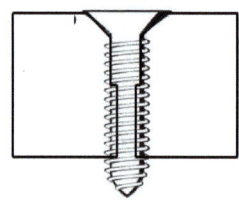

Preparing to place a screw

NAME	PURPOSE
Pin hammer	Used for driving pins/nails into the wood.
Lost head nail	Nail that can be hit below the surface of the wood.
Panel pin	Common pin for pinning.
Nail punch	Used for driving nails/pins under the surface of the wood, hiding them to improve a product's aesthetics.

KNOCK-DOWN FITTINGS AND FLAT-PACK FURNITURE

These fittings are perfect for the construction of flat-pack furniture. They are pre-made and designed to be easily fixed, simplifying furniture assembly. They are commonly used for joining manufactured boards. They can be bought in bulk, reducing the cost for the manufacturer while also making it easier to design the furniture, as they can be used in several different designs due to their adaptability. Flat-pack furniture further reduces the cost for the manufacturers, as they can:

- transport more of the item due to it being flat

- reduce the cost of manufacture because they do not need to assemble it

- improve sustainability due to reduced transport and manufacturing processes.

For the consumer, knock-down fittings mean they can easily assemble the product themselves, giving them a sense of achievement. Flat-pack furniture also reduces the end cost for the consumer due to the manufacturing cost reductions outlined above and the readiness of off-the-shelf furniture that can be fitted into your car for transport home.

 ONLINE

Learn more about knock-down fittings and the pros and cons of flat-pack furniture at www.brightredbooks.net

 ONLINE TEST

Test yourself on forming and joining wood at www.brightredbooks.net

 DON'T FORGET

Know the benefits of knock-down fittings and flat-packed furniture, as this is a popular question in the exam.

 ## THINGS TO DO AND THINK ABOUT

1. Describe the term 'knock-down fitting'.

2. Describe the advantages of flat-pack furniture for the consumer and the manufacturer.

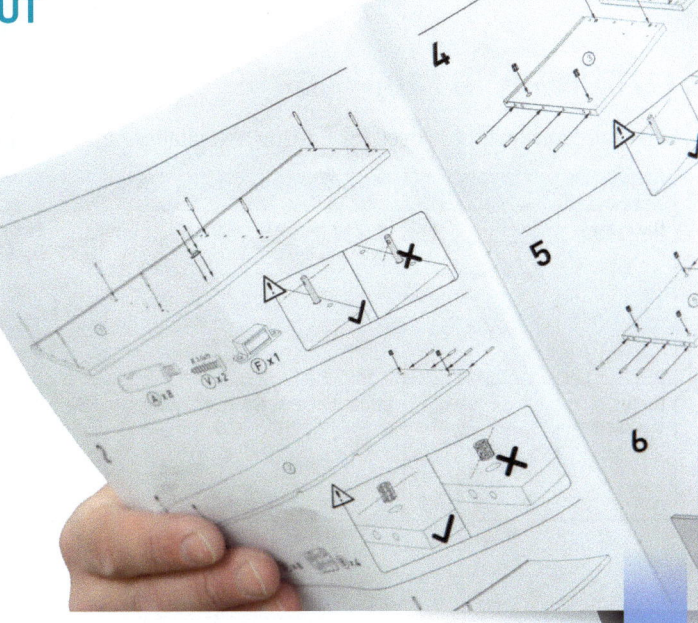

COMMON METALWORKING AND PLASTIC TOOLS

Engineer's vice

MANUFACTURING METAL

Metals are widely used in the manufacture of various products, and there are several processes that allow us to form them. At National 5 level, you must know the tools used to work metal and the following metal processes: shearing, notching, cold forming, hot forming, turning, die-casting and sand-casting. For each process, you should focus on learning the materials used for that process, the sort of products made by that process and the identifying features of the process.

There are various tools that can be used to cut, shape and form metal. It is important to note that many of these tools can be used in the manufacture of plastic as well. When working with metal or plastic, it is normally held in an engineer's vice.

MEASURING AND MARKING-OUT TOOLS

The following tools are used when marking out lines, shapes and patterns in metal and plastic. Steel rules are also used in the manufacture of metal and plastic.

TOOL NAME		PURPOSE	TOOL NAME		PURPOSE
Engineer's square		Like the try square, the engineer's square is used for marking straight lines at 90° to the edge of a piece of material. When using the engineer's square, you must ensure the handle is flat against the end of the metal/plastic.	Odd-leg callipers		The metalworker's version of a marking gauge. Used to mark parallel lines on the edge of a piece of metal.
Scriber		Basically a metalworker's pencil used to mark the metal.	Outside callipers		Used to measure the external diameter or width of metal/plastic rods and tubes. Commonly used when using a metal lathe, and can also be used for accurate measuring of diameters on a wood lathe.
Spring dividers		The metalworker's compasses, used for marking circles and arcs.	Inside callipers		As above, but for measuring internal diameters or widths.
Centre punch		Used to mark the centres of holes before drilling. Drills can slip on metal, and using a centre punch allows for more accurate drilling.	Micrometer		Highly accurate for measuring external diameters or widths. Used for ultimate precision when using a metal lathe, as they can measure to 0·01 of a millimetre.

CUTTING AND SHAPING

There are various tools used for cutting and shaping metal and plastic. The most common of these are listed below.

TOOL NAME		PURPOSE	TOOL NAME		PURPOSE
Hacksaw		Used for making straight cuts, or cutting metal or plastic to size.	Power hacksaw		An industrial version of the hacksaw that uses power to reduce the labour-intensiveness of cutting large metal rods or tubes.
Junior hacksaw		Similar to the hacksaw, the junior hacksaw is used for smaller jobs.	Rawhide mallet		A leather mallet used for shaping and forming metal. The soft leather ensures the mallet doesn't dent or damage the metal.

contd

Tin snips		Used for making straight cuts on thin sheets of metal.	Ball-peen hammer		Used for shaping and forming metal. The ball end of the hammer can be used to round and curve the metal.
Files		Files can be used to smoothen and shape metal or plastic. See pages 48–49 for more information on files and filing.	Cold chisel		Used for cutting metal or removing waste when other tools are not sufficient. Requires a lot of force and is normally used in conjunction with a forge (hot chisel).

CUTTING THREADS IN METAL

Threads can be cut in metal using split dies and taps. Threads are cut on the external surface of a metal rod or on the internal surface of a drilled hole.

External threads

To cut an external thread, a split die held in a die stock is used. The metal is held in an engineer's vice or metalwork lathe; and the split die, secured in the stock, is placed over the end of the metal before being turned in a clockwise motion to cut the thread. It is important that the die stock remains level and the metal is well greased to ensure a good cut.

Split die and stock

Internal threads

Internal threads are cut using taps. A **taper tap** with sloping sides, which makes the first cut easier, is used to start the cut. An **intermediate tap** is then used to cut deeper into the metal. Finally, a **plug tap**, which has straight sides and cutting teeth that cover the length of the tap, is used to finish the cut.

Taper tap in use

THINGS TO DO AND THINK ABOUT

1. To help prepare for your exam, you should begin producing rote answers that describe how to carry out cutting and shaping processes in metal/plastic with reference to the tools above.

2. Using YouTube, search for videos related to the above tools. This will help you to develop a better understanding of how they are used.

VIDEO LINK

Watch the clip on external threads at www.brightredbooks.net

VIDEO LINK

Learn more about internal threads at www.brightredbooks.net

DON'T FORGET

It is unlikely you that you will need to explain these tools in the exam, but you should learn what each tool's purpose is and be able to refer to the correct tool when explaining how to carry out processes in metal/plastic.

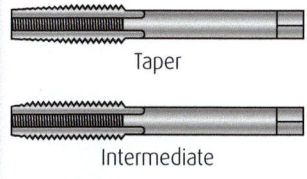

Taper

Intermediate

Plug

ONLINE TEST

Test your knowledge of common metalworking and plastic tools at www.brightredbooks.net

SHAPING AND FORMING METALS

CUTTING AND SHAPING

Metals can be shaped and formed using a variety of tools and processes. Hacksaws can be used to cut metal bars to length and cut out basic shapes in metal sheets. However, as metals can be difficult to saw, complex profiles in thin sheets of metal can be cut using **tin snips**. Tin snips act like scissors and are very useful for cutting complex curved profiles where it would be awkward to use a hacksaw. In addition to using hacksaws and tin snips, metals can also be cut in the following ways:

Shearing

Shearing is a process where metal is cut using a guillotine or tin snips. Shearing refers to taking a straight cut across the metal. Guillotines can be used to cut large metal sheets/bars to length.

Metal can deform if shearing is not carried out correctly. Deformation is where the metal bends, twists or fractures in the area where it is cut. Sheared metal can also be sharp, and therefore it must be filed down (see pp. 48–49).

Notching

The process of notching is carried out using a smaller guillotine or handheld notching tool, used specifically for cutting out notches and patterns in sheet metals. Deformation can also occur when notching, therefore it is important the process is carried out correctly.

FORMING

Forming refers to the process of shaping metal by bending, twisting and working the metal into a desired form. There are two ways to do this: **cold forming** and **hot forming**. In hot forming, the metal is heated to make it more easily bendable and to reduce the stress on the metal caused by forming. Metals are often formed around **formers** or **jigs** which are premade templates or forms that the metal can be bent around. *NOTE: formers and jigs can also be used in plastics.*

Cold forming

Folding: Metal sheets and bars can be folded using a number of processes. For smaller jobs, **bending/folding bars** secured in an engineer's vice can be used. The metal is placed in the bending bar before being bent over it using a rawhide mallet.

Folding machines can also be used. The metal is secured using a clamp on the folding machine. Levers are then pushed in an upwards movement to fold the metal to a desired angle. In industry, this is done using a powered folding machine.

Sinking/hollowing: Metal sheet is formed into a bowl shape using a mould/sandbag and bossing mallet. Bossing mallets are rounded, which forces the metal to curve.

Scrolling: Metal bars can be scrolled using a scrolling jig. Scrolling jigs are designed using a series of holes located in a spiral pattern. Scrolling pins can be placed in these holes, and the metal can then be forced around them to create a scrolled pattern.

Cold bending: Bending formers/jigs can be used to bend and form metal bars. This is done in a similar way to scrolling; however, the holes on the jig are laid out in a rectangular pattern. Pins can then be located in these holes as desired, and the metal can be formed around them.

Metal sheets can also be bent in this way; however, they are formed around blocks of wood/metal known as **formers** using a rawhide mallet rather than bending jigs.

DON'T FORGET

You must be able to explain how to mark out, cut and form metal. Make sure you have an awareness of the processes outlined on these pages.

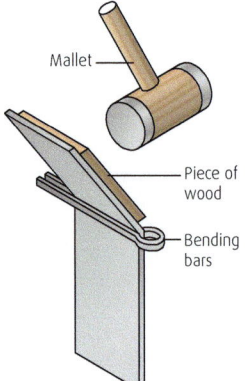

Mallet

Piece of wood

Bending bars

Bending bars in use

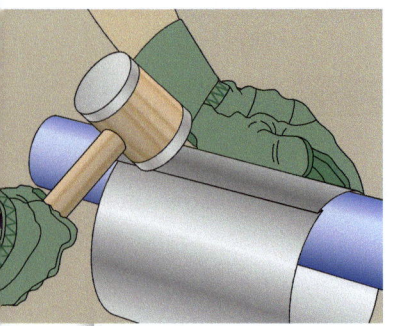

Forming a cylinder around a bar

VIDEO LINK

Check out the videos for this topic at www.brightredbooks.net

HOT FORMING

Many of the processes mentioned previously for cold forming can also be carried out when the metal is heated. However, heating the metal reduces the stress in the metal, softening it, which makes it easier to work.

THE FORGE

To heat the metal, a forge is used. Metal can be heated using the hearth or gas torch. The hearth allows you to heat a piece of metal and retain the heat, whereas the gas torch can be used to accurately heat a specific area of the metal. The forge is used in combination with an anvil, which provides a place to work the metal. As the metal is hot, it must be held using tongs.

Tongs and anvil in use

The forge

The following processes can be carried out on a forge:

Bending: The metal is heated up in the forge and can then be formed around a jig, the corners of the anvil or round bars secured in the anvil or engineer's vice.

Twisting: The metal is heated up in the area where the twist is to be formed. Once ready, the metal is held in an engineer's vice before being twisted using a twisting wrench.

Bending

Twisting

TEMPERATE HEAT TREATMENT

The properties of metal can be altered using heat treatment. This requires heating the metal to different temperatures before cooling it in different ways. Three common heat treatments include:

Annealing: This softens the metal, making it more malleable and easier to shape. Once heated to the correct temperature, the metal is left to cool naturally.

Hardening: This is the opposite of annealing. The metal is heated to just before critical (melting point). Once at this temperature, it is rapidly cooled in cold water. This causes the metal to harden. However, this process makes the metal brittle and difficult to work.

Tempering: This process is carried out after hardening. The metal can be heated to different temperatures, toughening it and reducing the brittleness of the metal. When working with steel, the temperatures used are as follows:

TEMP	COLOUR		USES
230°	Pale straw	HARDEST	Lathe tools and scribers
240°	Straw		Drill bits
250°	Dark straw		Taps, dies and centre punches
260°	Brown		Plane blades and lathe centres
270°	Brown/purple		Scissors and knives
280°	Purple		Cold chisels and saw blades
290°	Dark purple		Screwdrivers
300°	Blue	TOUGHEST	Springs and spanners

ONLINE TEST

Test yourself on shaping and forming metals at www.brightredbooks.net

DON'T FORGET

Make sure you can describe tin snips and bending bars, as well as how to use them.

DON'T FORGET

Make sure you can explain the processes of annealing, hardening and tempering.

THINGS TO DO AND THINK ABOUT

1. Describe how you could cut a curved profile using tin snips.
2. Describe the differences between annealing, hardening and tempering.

TURNING: CENTRE LATHE (METAL LATHE) 1

INTRODUCTION

Centre lathes are used for turning cylindrical forms of varying diameters from metal and plastic. Lathe work can be carried out manually or as a fully automated process using CNC lathes controlled by computers. The process can vary in cost due to labour and machine parts. If turning is carried out manually, it is only suited to low-volume production, whereas if it is automated it can be used for much higher volumes. When turning, it is important that the speed of lathe is correct for the size and type of material being turned, as well as the process being undertaken. This ensures a good-quality finish and limits any potential risks.

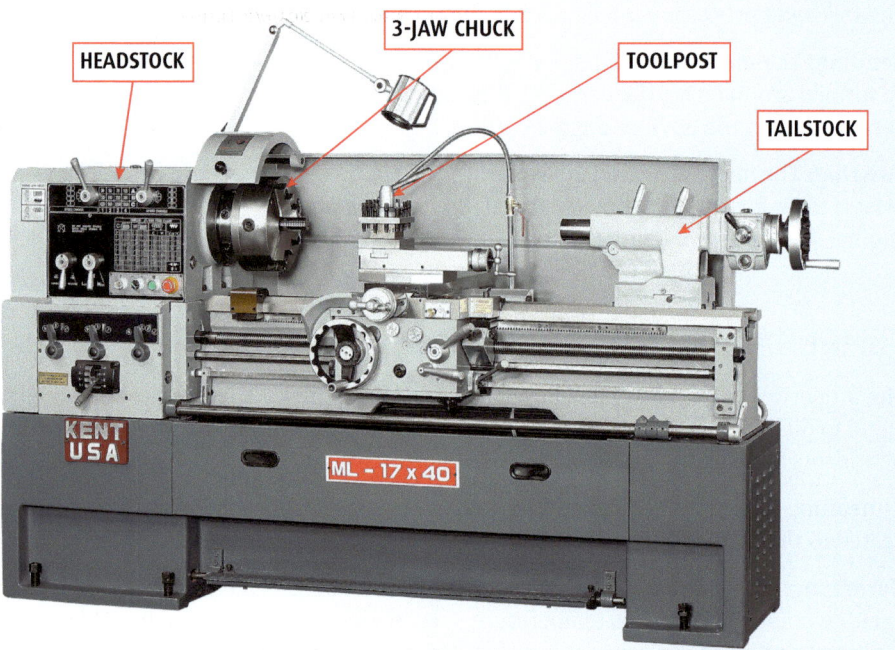

DON'T FORGET

You must learn how each of the processes of the metal lathe works. You may be asked to explain this in your exam.

LATHE PROCESSES

Cutting tools

There are a number of cutting tools that are used to carry out the various processes on a lathe. Each tool has a specific job. You do not need to explain these, but it will be useful to know what they are when trying to understand and describe each process.

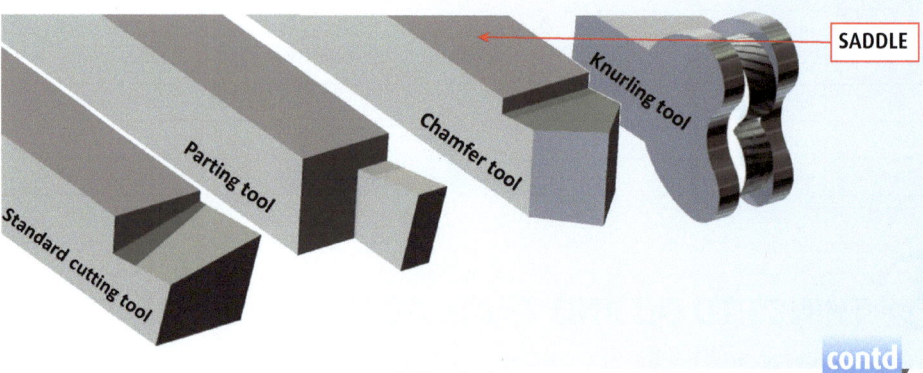

Cutting tools

contd

Facing off

Facing off refers to a cut being taken across the end of a piece of metal. This is done to remove any surface marks, flatten the surface to square it off and ensure the end of the piece has a high-quality finish.

Operation:

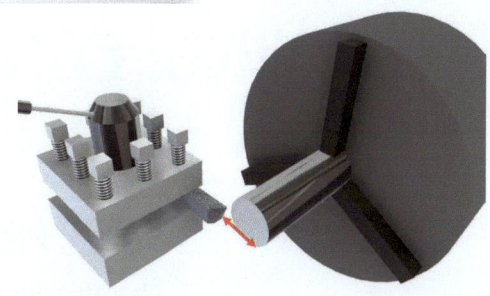

Facing off

- Set up a cutting tool in the tool post, and centre it vertically to the metal.

- Position the tool directly in front of the metal, and line it up with the end.

- Move the tool straight across the face of the metal, taking off no more than a few mm at a time.

Parallel turning and taper turning

Parallel turning refers to a parallel cut being taken along the length of a piece of metal to reduce its diameter. Taper turning is similar to this; however, rather than the cut being parallel, it is taken at a desired angle to create a slope along the length of metal.

Operation:

- Set up a facing/parallel cutting tool in the tool post, and centre it vertically to the metal.

- Position the tool at 90° to the end of the metal. If taper turning, rotate the tool post to the desired angle.

- Move the tool parallel along the surface of the metal, taking off no more than a few mm at a time, or along the metal at the desired angle for taper turning.

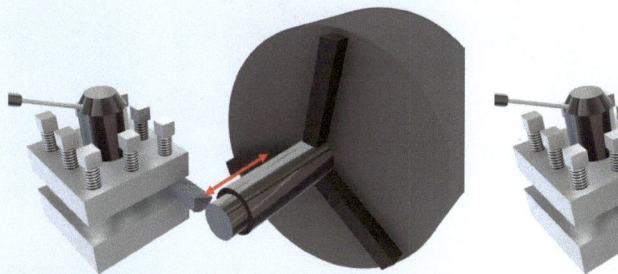

Parallel turning

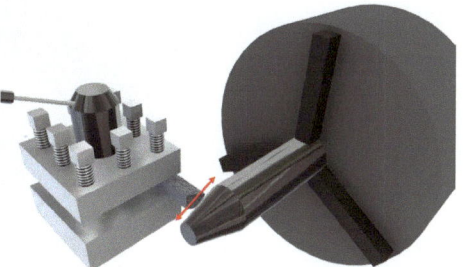

Taper turning

Step turning

Step turning uses the techniques described in parallel/taper turning to cut differing diameters into a piece of metal, producing a stepped effect in the metal. Each step can be as long as the manufacturer requires.

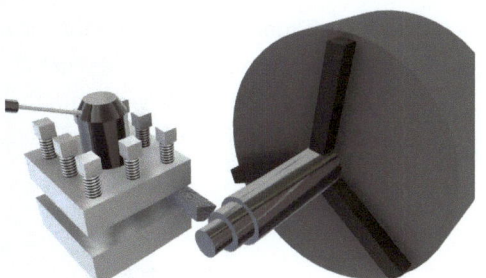

Step turning

VIDEO LINK

Check out the clip at www. brightredbooks.net to learn more about facing off.

VIDEO LINK

Check out the clip at www. brightredbooks.net to learn more about parallel and taper turning.

ONLINE TEST

Head to www. brightredbooks.net to test yourself on this topic.

THINGS TO DO AND THINK ABOUT

Prepare answers for each of the processes described on this page, using your own words. This will help you when answering these questions in the exam.

TURNING: CENTRE LATHE (METAL LATHE) 2

LATHE PROCESSES

Chamfering

Chamfering turns a square edge into a sloping edge. This can be done for aesthetic or for functional purposes.

Operation:

- A chamfer cutting tool is set up in the tool post and centred vertically to the metal.
- Chamfer cutting tools are usually cut to 45°.
- The tool is then directed into the edge of the metal, taking an angled cut.
- This can be repeated until the desired result is achieved.
- This process can also be carried out using a standard cutting tool and rotating the tool post to 45°.

VIDEO LINK

Check out the clip at www. brightredbooks.net to learn more about chamfering.

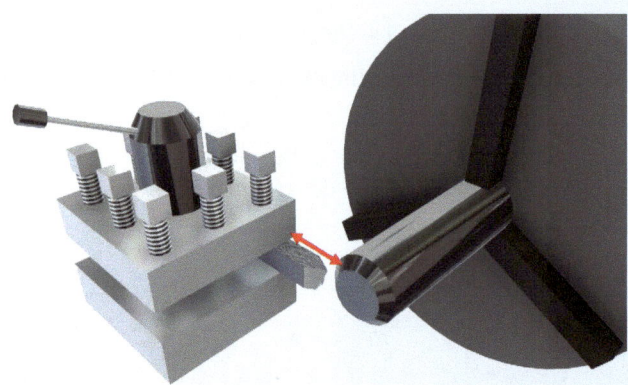

Chamfering

Parting off

Parting off allows the finished job to be removed from the excess metal.

Operation:

- A parting-off tool is set up in the tool post and centred vertically to the metal.
- The tool is positioned at 90° to the metal.
- The tool is then directed straight into the metal cutting.
- This is repeated until the required piece is freed from the excess metal.

VIDEO LINK

Check out the clip at www. brightredbooks.net to learn more about parting off.

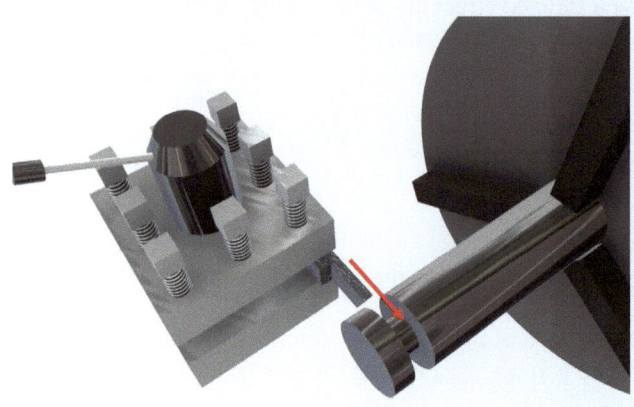

Parting off

contd

Knurling

Knurling creates a pattern on the surface of the metal. This is often carried out for functional reasons to provide a grip area on handles or bolts. It is important when knurling that the machine is set to a slow speed, as this will provide the best finish. Furthermore, the tool should remain in contact with the metal at all times until the operation is complete to ensure the knurl is cut correctly. Failure to do so will affect the end finish.

Operation:

- A knurling tool is set up in the tool post and centred vertically to the metal.
- The tool is positioned at 90° to the metal.
- The tool is then directed straight onto the surface of the metal.
- The tool is then moved parallel along the surface of the metal, creating the finish to the desired length, and is then reversed without leaving the metal.
- The depth of pattern is cut deeper by moving the tool further into the metal.

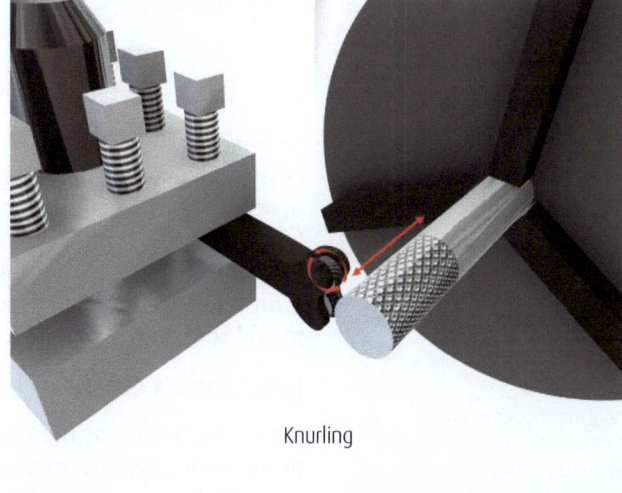

Knurling

Centre drilling and drilling

A centre drill is different from other drill bits (see pp. 82–83) in that its design reduces the chance of it skidding on the surface when drilling, due to the angle on the tip. It is used to accurately drill the centre before drilling a hole using standard drill bits. Centre drills and other drill bits are secured in a separate chuck (see 'Drilling', pp. 82–83) that can be connected to the tailstock. Once connected, the tailstock can be moved towards the spinning metal on the lathe. Turning the handle on the tailstock will move the drill bit forward/backwards, allowing you to begin drilling.

Centre drill

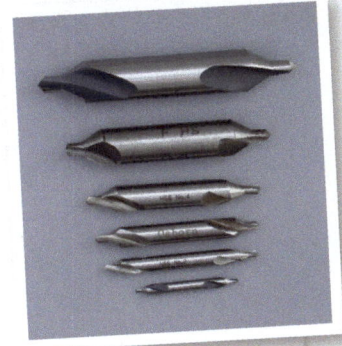

Materials used	Products made	Identification
Lathe work can be carried out using a variety of metals. Plastics can also be used, with nylon being the preferred choice.	Products made on a centre lathe include: nuts, bolts, tool shafts, turned chess pieces and decorative features such as the end caps on curtain rails.	Products turned on a centre lathe will be cylindrical.

THINGS TO DO AND THINK ABOUT

1. Prepare answers for each of the processes described on this page, using your own words. This will help you when answering these questions in the exam.

2. Watch the videos at www.brightredbooks.net

   Each video shows metal being turned on a centre lathe. Watch these, and identify each of the lathe processes being carried out. This will further enhance your understanding of lathe work and allow you to see how these processes can be combined when making one component.

 VIDEO LINK

Check out the clip at www. brightredbooks.net to learn more about knurling.

 VIDEO LINK

Check out the clip at www. brightredbooks.net to learn more about centre drilling.

 ONLINE TEST

Head to www. brightredbooks.net to test yourself on this topic.

DON'T FORGET

When answering questions on each process, break your answer down into steps as is done in the bulleted descriptions for every lathe processes given.

METAL PROCESSES IN INDUSTRY

DIE-CASTING

In metalwork, die-casting is similar to the plastic process of injection moulding. Die-casting allows manufacturers to form metal into complex designs with a high degree of accuracy. The end product requires very little finishing, where only the sprue feed (channel through which the metal is injected) and any flashes (leaked material) need to be removed. It is a highly automated process that is expensive to run, and therefore it is only suitable for mass production, due to high set-up costs, high tooling costs, high mould costs and maintenance costs.

How does it work?

A chamber known as the 'charge chamber' is filled with molten metal. An injection piston then forces the molten metal into a pre-designed split mould. A split mould is a mould that comes in two halves. Once in the mould, the metal is then cooled within, using cold water. Once the metal solidifies, it is released from the mould, and any finishing work is then carried out.

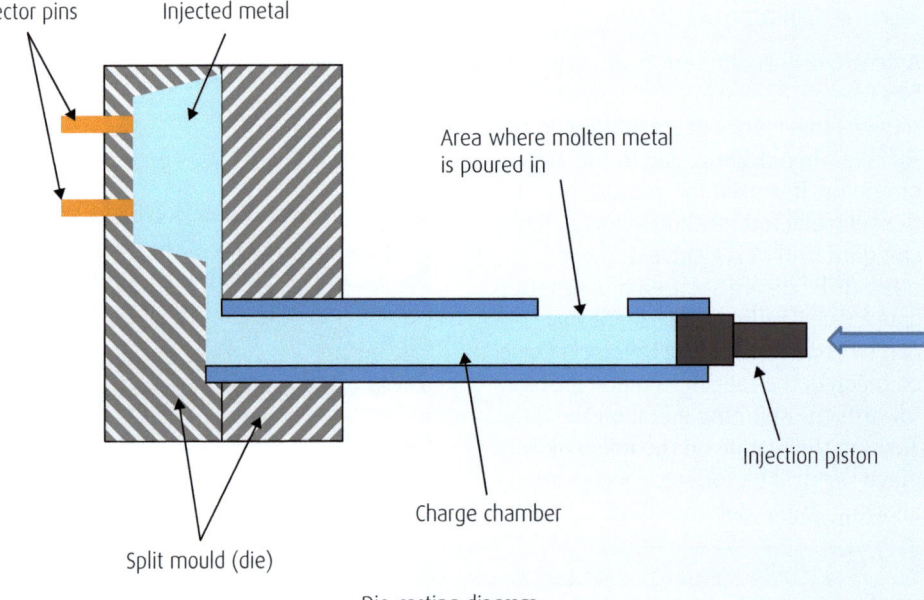

Ejector pins   Injected metal

Area where molten metal is poured in

Injection piston

Charge chamber

Split mould (die)

Die-casting diagram

Materials used

The most commonly used materials in die-casting are metals and alloys that have a low melting point, such as zinc, aluminium, magnesium and brass.

Products made

The products made using die-casting include toy cars, engine parts, alloy wheels, tools, and metal casings for electronics such as disc drives and mobile phones.

Identification

Like injection moulding in plastic, die-casting is highly accurate and can form very complex shapes with a high-quality end finish. Different surface finishes and textures can also be achieved. Other features such as tapered edges, ejection pin marks, sprue marks, runner marks and mould split lines can be found on die-cast products, making them easier to identify.

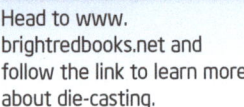

ONLINE

Head to www.brightredbooks.net and follow the link to learn more about die-casting.

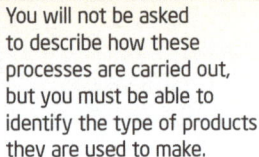

DON'T FORGET

You will not be asked to describe how these processes are carried out, but you must be able to identify the type of products they are used to make.

VIDEO LINK

Check out the clip at www.brightredbooks.net to learn more about die-casting.

SAND-CASTING

Sand-casting is a traditional process that is the most widely used casting process in industry. However, its labour-intensive nature makes it more suited to batch production. Products can vary from small to large in size. Costs can also vary depending on the set-up. An automated system will obviously cost more than a manual one.

How does it work?

Place a mould into the drag (bottom half), and tightly pack sand around it. Next, attach the cope (top half) to the drag. Insert a runner and riser (sprue feeds), and tightly pack sand around them. Separate the cope and drag, and then remove the mould, runner and riser. Re-attach the cope and drag. Pour molten metal into the runner. Once the molten metal comes out of the riser, you know the mould is filled. The mould is then left to cool, and once it is solidified it is removed. When the finished product is removed, the metal formed in the runner and riser will still be attached. This should be removed using a hacksaw before any final finishing work is carried out.

CRUCIBLE:
Metal is melted in the crucible and poured from this

POURING BASIN:
Wider hole at the top to prevent the molten metal from spilling over the edge

RUNNER:
Molten metal is poured in here

COPE:
Top half of the moulding box

DRAG:
Bottom half of the moulding box

RISER:
Molten metal exits here once the mould is full

MOULD:
The hollow shaped mould for the product being made

Sand-casting diagram

Materials used

Most metals can be used in sand-casting, although metals with a low melting point are preferable. Most products made using sand-casting are made from iron, aluminium, bronze and brass.

Products made

Casting is a versatile process, and products made include plumbing fittings, engine parts, pump housings and tools.

Identification

Products made from sand-casting can be rough in appearance initially, and they require a lot of finishing, ensuring that a high-quality surface finish is achieved. To make it easier to remove the finished products from the mould, draft angles, rounded corners and tapered edges are used. These are often evident on the end product. Finally, there are often marks where the runner and riser have been removed.

 ONLINE

Head to www.brightredbooks.net and follow the link to learn more about sand-casting.

VIDEO LINK

Watch the clips at www.brightredbooks.net to find out more about sand-casting.

 THINGS TO DO AND THINK ABOUT

ONLINE TEST

Test your knowledge of casting at www.brightredbooks.net

1. Carry out some of your own research into these processes, and compare the advantages and disadvantages of each for the manufacturer. You should make a note of these and include it in your revision notes.

2. Research other products that are made using sand-casting and die-casting. This will give you a better understanding of how and where these processes are used.

JOINING METALS

There are several ways of joining metal. It is important to select the correct method for use in a product to ensure it is structurally sound. You should learn the following metal-joining techniques.

RIVETS

Rivets are used to permanently join metal together. They can be used in small or large projects such as buildings and bridges. There are two methods for joining metal using rivets:

1. Solid rivet

This method uses rivets that require hammering to fix the metal together. There are four types of rivet used in this method:

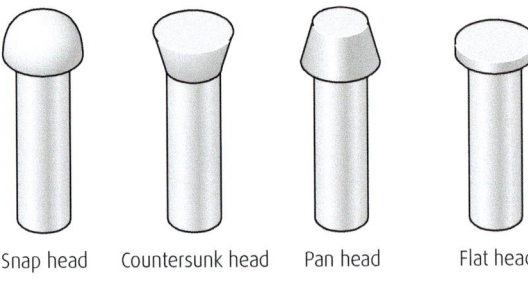

Snap head   Countersunk head   Pan head   Flat head

Each of the rivets produces a different finish. For example, a countersunk rivet will sit level with the surface of the metal, whereas a snap head will leave a half-dome finish above the surface of the metal.

To form a rivet joint, a rivet is pushed through a matching hole in two pieces of metal. Using a rivet set, which protects the shape of the chosen rivet, the metal is held firmly together, and a ball-peen hammer is used to form the other end of the rivet to fix the metal together. The end of the rivet can be flattened or formed to match the type of rivet in use.

Ball-peen hammer
(relative size reduced)

Rivet ——

Start of
—— rolled edge

Rivet set

Rivet being fixed

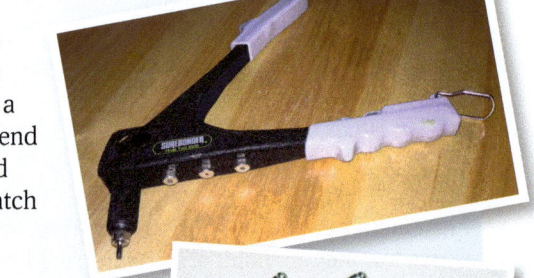

2. Pop rivet

Pop rivets are much quicker than solid rivets. They are fixed using a pop rivet gun and a special pop rivet. Again, the pop rivet is pushed through two matching holes in two pieces of metal. A pop rivet gun is then placed over the pop rivet. The handle (trigger) is squeezed, and the rivet expands in the hole until it is fixed and the mandrill pin pops off.

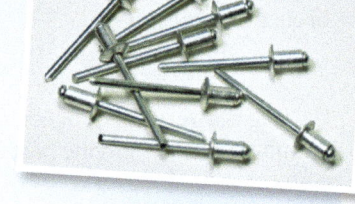

Pop rivets and pop rivet gun

VIDEO LINK

Learn more about this by watching the clip at www.brightredbooks.net

WELDING

Welding forms a solid joint where a filler material is melted and then solidified between two pieces of metal, fusing them together. There are three types of welding you should know:

1. MIG welding

MIG welding uses a reel of filler material that is continuously fed through a welding gun. The filler material is melted as it comes out of the gun, allowing the user to run a line of the filler material, known as a **bead**, between the two pieces of metal being joined. This process is much easier than TIG welding.

VIDEO LINK

See how a pop rivet works by watching the video at www.brightredbooks.net

VIDEO LINK

Check out the clip at www.brightredbooks.net to learn more.

contd

2. TIG welding

TIG welding works in two ways. **Gas welding** uses a gas-flame torch to melt a separate rod of filler material. **Arc welding** passes an electric current between the filler rod and the metal, creating a current that melts the filler rod. This is a delicate process and is better suited to smaller jobs.

3. Spot welding

Two pieces of metal are placed together in between the copper rods of a spot welder. The copper rods are then closed over the metal, passing an electrical current through them and the metal, which fuses the two pieces of metal together.

Spot welding

SOLDERING

Soldering takes two forms: **soft soldering** and **hard soldering** (brazing).

Soft soldering requires the use of a soldering iron. A filler material called solder, made from a mixture of tin and lead, is melted using the heat of the iron, which forms a solid join. It is ideal for small jobs; and soldering irons can be manual, where the tip must be heated using a forge, or they can be electric. Manual soldering irons can be difficult to use, therefore they are used in jobs where you have space to work to ensure a good finish. Electric irons are commonly used for soldering circuits due to their design, which ensures higher accuracy.

Soft-soldering circuits using an electric soldering iron

Hard soldering (brazing) uses a silver solder filler material mixed from copper, zinc and silver. It can also be referred to as **silver soldering** for this reason. Unlike soft soldering, it requires the use of a torch to melt the solder due to its high melting point. It is commonly used in plumbing work to permanently join pipes together.

Brazing copper pipes

Bolts and machine screws

Nuts and bolts provide a non-permanent fixing whereby a bolt is pushed through two holes and a nut is used to tighten the metal together. **Machine screws** are made to fit the insides of pre-threaded holes in metal.

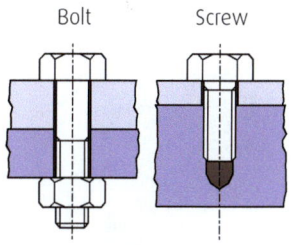

Difference between a bolt and machine screw

GLUING

Metal can be glued together; however, this is not a particularly strong way to join metal in jobs that require good strength properties. One of the most common glues used for metal is **epoxy resin**. Epoxy resins normally come in two parts that require mixing: a hardener and the resin. Once mixed, these react and then set, creating a strong bond.

THINGS TO DO AND THINK ABOUT

1. Describe the three types of welding used to join metal.
2. Describe the difference between soft and hard soldering.

VIDEO LINK

Watch an example at www. brightredbooks.net

VIDEO LINK

Watch this process at www. brightredbooks.net

ONLINE

http://www.bbc.co.uk/ schools/gcsebitesize/design/ resistantmaterials/jointsrev4. shtml

DON'T FORGET

Make sure you know these methods, as well as their most appropriate applications when joining metals.

SHAPING, FORMING AND JOINING PLASTICS

As outlined previously, plastics are highly versatile and can be formed, moulded and shaped easily.

SHAPING

Plastics can be easily shaped using files to round corners and to form curves, notches or patterns in the plastic. Using the process of cross-filing (see p. 46), the file can be rotated side-to-side to round corners. Using differently shaped files can allow you to cut square notches or curves into the plastic. One of the easiest ways to shape plastic is cutting. Plastic can be sawn to shape using a coping saw or bandsaw; however, it is important that the plastic is properly secured, as some plastics snap easily.

VIDEO LINK

Check out the clip on line bending in industry at www. brightredbooks.net

FORMING

To form plastic, it must be heated until it softens. Thermoplastics, remember, are able to return to their original shape (plastic memory), which allows us to change their form several times if we are not happy with the finished outcome. Once softened, plastic is extremely flexible, therefore moulds can be used to make forming them easier and more accurate. To form plastic, we can use two processes:

1. Line bending

Line bending is used to bend thin sheets/strips of plastic and requires a machine called a strip heater. Strip heaters heat the plastic along an exact line marked on the plastic. Only one bend can be made at a time, therefore the process is quite time-consuming.

Strip heater

2. Hot forming

To twist, mould or curve plastic, we can use a plastic oven to heat it and then form it around a mould or jig. The thickness of the plastic will determine the temperature and length of time the plastic will need to be heated for. Once heated, the plastic should be set around the mould and left to cool.

JOINING PLASTIC

As plastics can be moulded in one complete shape or product, they often do not need to be joined. However, where they do need to be joined, the following techniques can be used.

Gluing

Plastic can be glued together using **Tensol plastic cement** or **epoxy resin**. Both of these are very strong contact adhesives and fuse the plastic together. Epoxy resin can also be used to help join metal and plastic together, but wood is more difficult to join to plastic.

Plastic rivets

Plastic rivets provide a semi-permanent fixing that joins the plastic together. There are two main types of rivet. The **ratchet rivet** uses 'male' and 'female' parts that snap together when pushed through matching holes on two pieces of plastic. **Snap/fastener rivets** grip the plastic when pushed through a hole using specially designed teeth located on the outside of the rivet that stop the plastic from separating.

Plastic oven

VIDEO LINK

For more on this, check out the clip at www. brightredbooks.net

DON'T FORGET

Make sure you learn the processes of hot forming and line bending as well as the machinery required for each.

contd

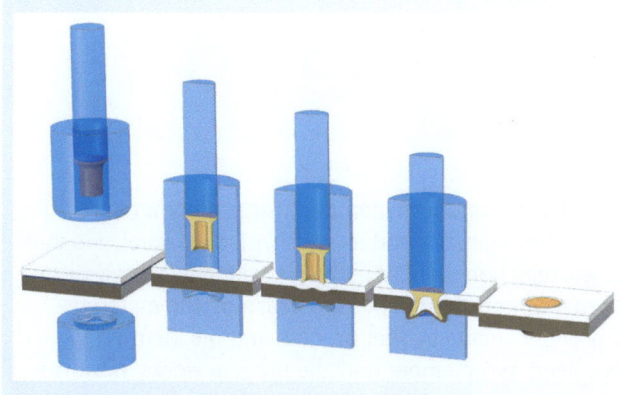

Snap/fastener rivet diagram

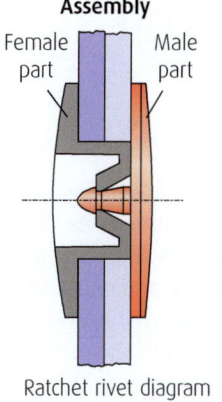

Assembly

Female part / Male part

Ratchet rivet diagram

PLASTIC WELDING

There are various forms of plastic welding. The aim of welding is to fuse two pieces of plastic together by melting them along with a filler material that seals the gap. Two of the main welding techniques are extrusion welding and speed tip welding.

Extrusion welding

Extrusion welding is the preferred technique for joining material over 6 mm thick. Thin plastic rod is drawn into a miniature handheld plastic extruder, melted, and forced out of the extruder against the parts being joined, which are softened with a jet of hot air to allow bonding to take place.

Speed tip welding

A plastic welder, similar to a soldering iron, is fitted with a feed tube for plastic rod. The speed tip heats the rod at the same time as it presses the molten rod into position. A bead of softened plastic is laid into the area where the plastic is to be joined, and the melted rod fuses them together.

JOINING PLASTICS TO WOOD AND METAL

Joining plastics with wood and metal can be difficult. Strong contact adhesives such as **epoxy resin** work with certain materials. An easier way is to use screws when joining plastic to wood. **Bolts and nuts** can also be used for joining plastic to wood and metal. See pages 74 and 75 for more information on these processes.

VIDEO LINK

Watch the video at www. brightredbooks.net for more on plastic welding.

DON'T FORGET

You may not be asked directly about plastic joining processes in the exam, but you are expected to have an awareness of these.

THINGS TO DO AND THINK ABOUT

1. Describe how you would form the base part of the plastic hook shown here using the process of line bending.

2. Describe how the plastic drinking straw shown here could be formed.

PLASTIC PROCESSES IN INDUSTRY 1

INTRODUCTION

Plastics are the most widely used material in modern manufacturing, and there are several processes that can be used to form plastic products. At National 5 level, you must know the following plastic processes: injection moulding, rotational moulding, vacuum forming and extrusion. For each process, you should focus on learning the materials used for that process, the sort of products made by that process and the identifying features of the process. You do not need to remember how the process works; however, information on this is provided over the next few pages to improve your understanding.

DON'T FORGET

For each plastic process, learn the materials used, products made and identifying features.

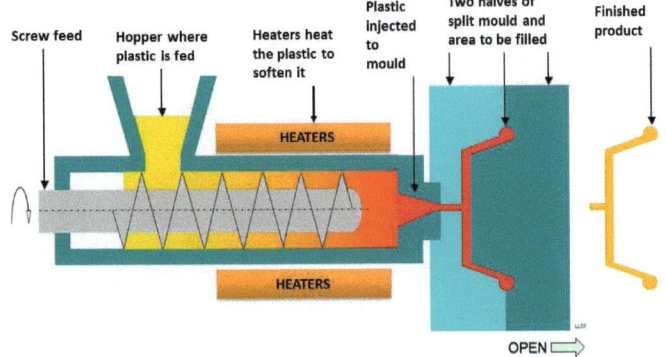

Injection-moulding diagram

INJECTION MOULDING

Injection moulding is one of the most commonly used processes in the manufacture of plastics. It is a highly automated, high-precision process that allows identical products to be made quickly and in high volume. Injection moulding is only cost-effective under mass production, being expensive to run due to the set-up costs, high tooling costs, high mould costs and maintenance costs. Therefore companies will only recover their manufacturing costs if they have large quantities of their product to sell.

How does it work?

A hopper is filled with thermoplastic powder or pellets. A rotating screw then passes the plastic through a heater, where it plasticises (softens). The softened plastic is then injected into a mould which has been designed to form the shape of the final product. Once injected fully, the plastic is left to cool and set. Finally, once the plastic has set, the finished product is ejected from the mould.

Materials used

Only thermoplastics are used in injection moulding. The most popular choices are ABS, polypropylene, polythene, polystyrene and nylon. Thermosetting plastics are **not** used because they would harden inside the machine, causing damage which would be very expensive to repair.

Products made

The products made using injection moulding range from small items such as toothbrushes, bottle lids and plastic cutlery to larger items such as plastic casings for various electronic products, ski boots, tables and chairs.

Identification

Injection moulding is highly accurate and can form very complex shapes with a high-quality finish. Different surface finishes and textures can also be achieved, such as the grip detail on a bottle cap. Other features such as draw angles, ejection pin marks, sprue marks and mould split lines can be found on injection-moulded products, making them easier to identify.

ONLINE

Learn more about injection moulding at www.brightredbooks.net

VIDEO LINK

Check out the clip at www.brightredbooks.net to learn more about injection moulding.

ROTATIONAL MOULDING

Rotational moulding is used to produce hollow plastic products that vary in size. Rotational moulding is a highly automated process that is versatile and relatively inexpensive. Due to its low production rate, it is limited to batch production.

contd

How does it work?

Plastic powder is added to a mould. The mould is then closed and heated. The mould begins to rotate on two axes. As the plastic powder comes into contact with the mould surface, it melts and the rotation causes the molten plastic to evenly coat all surfaces of the mould. The mould's surface then begins to cool, and the plastic sets. Once set, the finished product is removed from the mould.

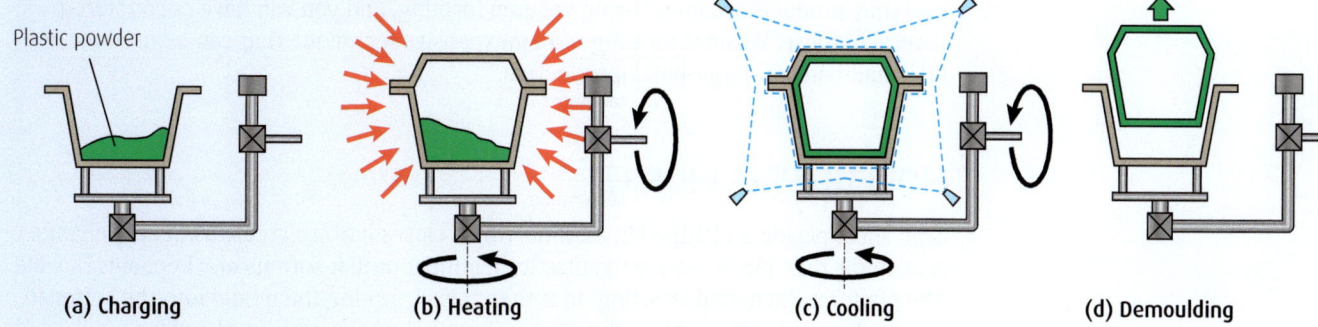

Plastic powder

(a) Charging   **(b) Heating**   **(c) Cooling**   **(d) Demoulding**

Rotational moulding diagram

Materials used

Thermoplastics, mainly low-density polythene, as well as polypropylene and nylon, are used in rotational moulding. However, some thermosetting plastics can also be used.

Products made

Typical products made by rotational moulding can include kayaks, canoes, dustbins, storage containers, water tanks, traffic cones, children's playhouses and plastic garden sheds.

Rotational moulded product

Identification

Rotational moulded products are hollow and have a uniform wall thickness throughout. They also have a high-quality surface finish; and details such as logos, text and information graphics can be imprinted onto the surface during the process.

Logo imprints

ONLINE

Head to www.brightredbooks.net and follow the link to learn more about rotational moulding.

VIDEO LINK

Watch the clip at www.brightredbooks.net to find out more about rotational moulding.

ONLINE TEST

Test your knowledge of injection and rotational moulding at www.brightredbooks.net

THINGS TO DO AND THINK ABOUT

1. To further develop your understanding of plastic processes, you should begin making revision notes that will help you study for your exam. Draw up a four-column table with the headings: Process, Materials, Products and Identification. Under each heading, condense the information above into your own words, and use your table for quick revision of each plastic process covered in this book.

2. Look around your home and try to identify any plastic products that you think have been made using injection or rotational moulding. Write down your reasons for choosing each product, and discuss these in class with your teacher and classmates.

PLASTIC PROCESSES IN INDUSTRY 2

VACUUM FORMING

Vacuum forming is one the oldest techniques used in the manufacture of plastics. Many everyday products are made using vacuum forming, and you will have encountered several of these. Vacuum forming is an inexpensive technique that can be used for one-off, small-batch or large-batch production.

HOW DOES IT WORK?

Thin sheet plastic is clamped in a frame, which closes it off to create a vacuum chamber. A heater is then placed above the plastic, heating it until it softens and becomes flexible. The platform the mould is sitting on is then raised, forcing the mould into the softened plastic. The air in the vacuum chamber is then sucked out, causing the plastic to form tightly around the mould. The plastic is then left to cool and set to form the end product. Finally, any excess material is removed.

DON'T FORGET

Remember why vacuum-forming moulds have tapered edges/rounded corners, and be able to describe thinning.

1 Plastic is secured and then heated.

2 Once softened, the platform is raised, forcing the mould into the plastic, and the air is removed, forming the plastic tightly around the mould.

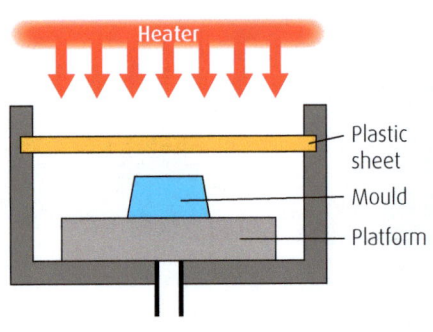

Plastic sheet

Mould

Platform

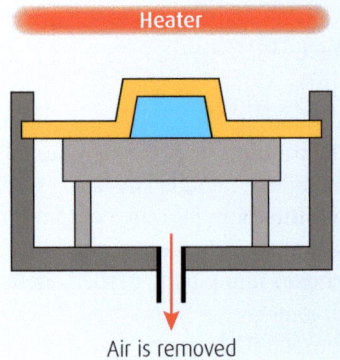

Air is removed

Vacuum-forming diagram

Materials used

Only thermoplastics are used in vacuum forming. The most popular choices are ABS, acrylic, polythene, high-density polystyrene and PVC. Unlike some other processes, vacuum forming requires some final processing where excess material has to be trimmed off. This excess material can be ground down and recycled.

Products made

The products made using vacuum forming include: dinner trays, yoghurt pots, food containers, sandwich boxes, chocolate trays, ice-cube trays, lighting panels and shower trays.

Identification

Vacuum forming requires the use of thin sheet plastic, and therefore vacuum-formed products can be easily identified by their thin wall thickness. Furthermore, any patterns or textures found on the mould will be imprinted on the end product. Due to how tightly the plastic is formed around the mould, vacuum-formed products will have **tapered** (sloped) edges and **rounded** corners that make it easier to remove the product from the mould once set. Lastly, because the plastic is stretched tightly over the mould, the corners are subject to **thinning**. Thinning literally means the plastic has thinned in the areas where it has been subject to the extremes of stretching over the mould.

ONLINE

Learn more about vacuum forming at www.brightredbooks.net

VIDEO LINK

Check out the clip at www.brightredbooks.net to learn more about vacuum forming.

EXTRUSION

Extrusion is a process that can be used in both plastic and metal manufacture. It is used to produce continuous lengths of a cross-section. To help you understand this better, think about a round pipe. A pipe can come in various lengths, and the cross-section is its end shape – a circle. Extrusion is best used for high-volume batches, as machine parts and dies (moulds) can be quite expensive.

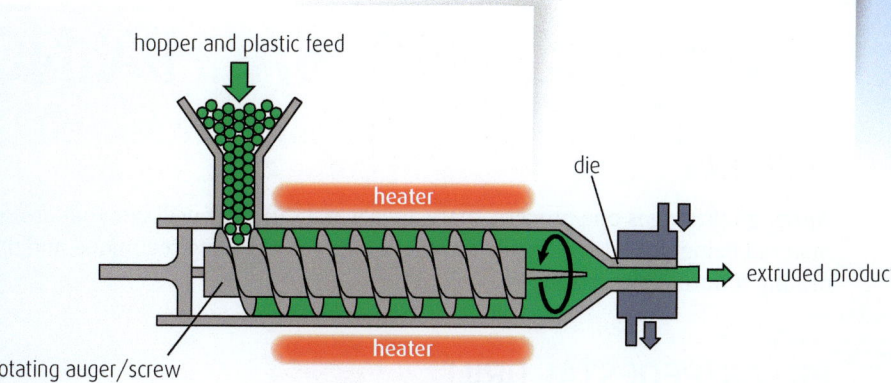

hopper and plastic feed

heater

die

extruded product

heater

rotating auger/screw

Plastic-extrusion diagram

How does it work?

Plastic extrusion works in a similar fashion to injection moulding. A hopper is filled with thermoplastic powder or pellets. A rotating screw then passes the plastic through a heater, where it plasticises (softens). The main difference happens here. Instead of the plastic being injected into a mould, it is extruded through a die designed to produce the cross-sectional shape of the end product. The extruded product

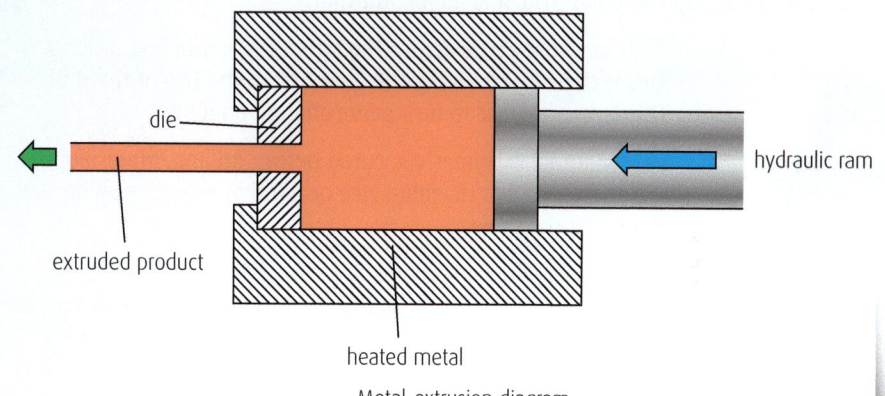

die

extruded product

hydraulic ram

heated metal

Metal-extrusion diagram

then passes through a cooling section so that it can harden and set. In **metal extrusion**, heated metal is forced through the die using a ram instead of the rotating screw method described above.

Materials used

Thermoplastics such as polythene, PVC and polypropylene are suitable for plastic extrusion. Metals such as lead, copper, aluminium, brass and steel are suitable for metal extrusion. It is important to note that plastics are considerably easier to extrude than metals, as less force is required.

Products made

The products that can be made using extrusion include: curtain rails, plastic and metal pipes, guttering, light covers, hosepipes, drainpipes, electric cable coverings and PVC window/door frames.

Extruded products

Identification

Due to the nature of extrusion, products will have a uniform wall thickness along the entire length of the product. Furthermore, the cross-section (end shape) will be uniform and unchanged along the length of the product. Finally, accuracy is very high, and the shapes produced can vary in complexity.

ONLINE

Head to www. brightredbooks.net and follow the link to learn more about extrusion.

VIDEO LINK

Watch the clips at www. brightredbooks.net to find out more about rotational moulding.

DON'T FORGET

You do not need to know extrusion for the exam; however, it is a useful process to be aware of, and will help you when completing coursework.

THINGS TO DO AND THINK ABOUT

1. Explain why moulds for vacuum forming have rounded corners and tapered sides.

2. Describe the term 'thinning' when vacuum forming.

ONLINE TEST

Test your knowledge of vacuum forming and extrusion at www. brightredbooks.net

DRILLING AND MACHINING

DRILLING

Although drilling is one process, there are several different drill bits that are required for the type of hole and material being drilled. It is important that the correct drilling technique and drilling equipment are used when drilling different materials.

PILLAR/PEDESTAL DRILL

This is the standard workshop power drill. The table allows for accurate, level drilling. To set up the pillar drill, it is important that:

Jacobs chuck

- The correct speed is set for the type of material and drilling. This is done by changing the gear belt at the top of the drill. This is very similar to how gears change on a bike.

- Your material is properly secured before drilling. Materials are often held in a machine vice or hand vice before being placed on the table.

- You are using the correct drill bit for the job and the drill bit is properly secured in the Jacobs chuck. A Jacobs chuck is a three jaw chuck that requires tightening with a chuck key. The three jaws ensure the drill bit stays true (spins uniformly). It is important that the chuck key has not been left in the Jacobs chuck before starting the drill.

- Set the depth stop where you are not planning to drill all the way through material.

Location of Gears
On/off
Safety guard
Pillar

Control lever for controlling the drilling depth

Location of Jacobs chuck

Drilling table

Pillar/pedestal drill

Hand vice

Machine vice

Setting up the drill depth stop: The depth stop can be used to control the depth of drilling. This is useful when drilling holes to a specific depth e.g. for creating a dowel joint. The depth stop is located on the lever. A lock is tightened and the measuring gauge is used to set the depth.

Other types of drill and drill bits

DRILL		PURPOSE
Handheld power drill		These can be battery-operated or powered using a cable. Power drills ultimately make drilling easier and also have the advantage of being easily portable.
Handheld drill		A manual drill operated by winding a handle to turn the drill bit.
Bit-and-brace drill		For manual drilling. The brace allows the user to apply pressure by using their weight and leaning into the drill. The wider handle allows more force to be applied, which is good for drilling bigger jobs.

DRILL		PURPOSE
Twist bit		Standard drill bit for all materials. Twist drills come in various forms and sizes. Hardened bits can be used for drilling tough metals.
Fostner bit		Used for drilling flat-bottomed holes in wood.
Auger bit		Used in hand drills. Makes drilling easier due to the widened twists on the bit, which removes more material.
Flat bit		For drilling larger holes and flat-bottomed holes.
Countersunk bit		Used for countersinking the tops of holes to allow countersunk screws to sit below the level of the wood's surface.
Hole saw		Used to drill large holes or for cutting discs out of various materials.

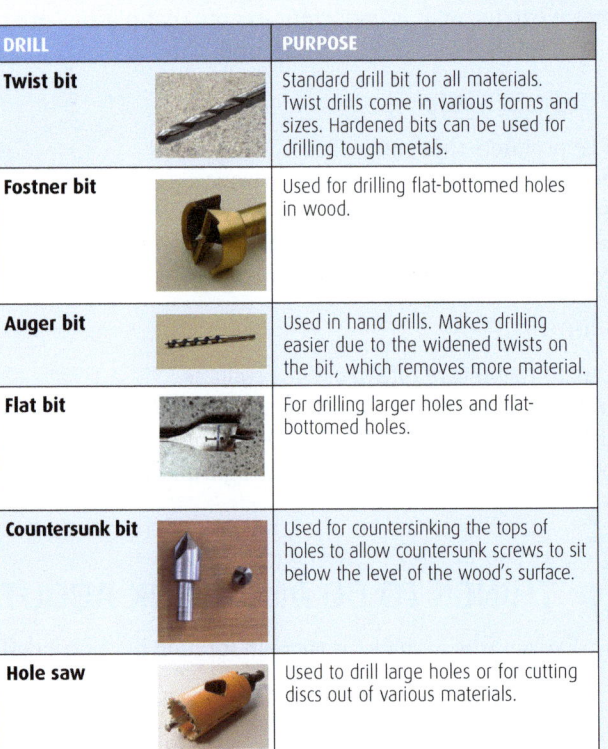

DON'T FORGET

Know the how to set up the pillar drill and how to set the depth stop.

MORTISE MACHINE

The mortise machine works like a pillar drill that cuts square holes. It is most often used for cutting out the mortise for a mortise and tenon joint. It cuts using a hollow square chisel that houses a drill bit. The drill bit drills into the wood to remove waste whilst the chisel squares off the hole. It can also be used for cutting large square holes into woods.

Mortise bit

Setting up the mortise machine

- Ensure the wood is secured in the table clamp.
- Ensure the correct size of mortise bit is set up for the cut being taken.
- Correctly position the material under the mortise bit using the table adjustment wheels.
- Set the depth stop to the required depth for the cut.

Setting the mortise depth stop

The depth stop on the mortise machine can be set by locking the handle on the diagram shown here. The handle locks a bar that stops the mortise bit going further than intended.

Often the easiest way to set the depth stop is to lower the mortise bit towards a desired depth marked on the wood. Once set against this the depth stop can be locked.

SANDING MACHINES

Sanding machines are used for preparing and shaping woods. There are two types of sander you must be familiar with:

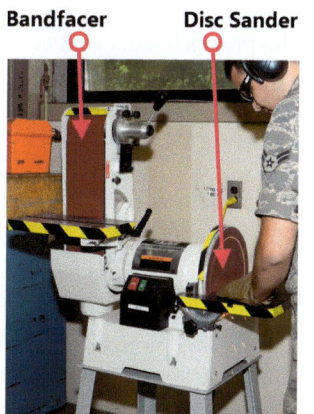
Bandfacer   **Disc Sander**

- **Disc Sander (and band facer):** These are useful when squaring the ends of wood after sawing, creating curved corners on wooden strips or boards, and shaping a curved profile. The disc sander works via a rotating disc covered in abrasive paper, and the band facer works via a rotating belt covered in abrasive paper. When using either machine it is important your hands are kept behind the material at all times and the material remains flat to avoid risk of injury.

- **Belt Sander:** The belt sander is an automated handheld power sander used for smoothening and finishing the surface of wood. This is much easier than manually hand sanding wood and is particularly useful for large jobs. Care has to be taken however, as belt sanders are extremely powerful. Failure to monitor the amount of sanding taking place can result in too much material being removed. In addition to this, belt sanding veneered manufactured boards can often lead to the veneer being stripped off, meaning due care must be taken.

ONLINE

Learn more about drill bits at www.brightredbooks.net

VIDEO LINK

Check out the clip at www.brightredbooks.net for more.

DON'T FORGET

Know the uses and set up for the machines on this page.

ONLINE

Find out how to drill a hole step by step at www.brightredbooks.net

ONLINE TEST

Test your knowledge of drilling at www.brightredbooks.net

THINGS TO DO AND THINK ABOUT

Make your own notes on how you would set up and drill a hole in wood, metal and plastic using a standard twist drill. This will help you in preparing for your exam.

MANUFACTURING IN INDUSTRY AND SOCIETY

CAM AND COMMERCIAL MANUFACTURE

INTRODUCTION

Commercial manufacture refers to products being made in an industrial scale/setting using specialist processes from hand manufacture to industrial machinery such as injection moulding. Products in manufacture often fall into one of four categories:

1. One-off production

Only one product is made at a time. Every product will be unique, and the manufacturing process is often very labour-intensive, as it will require skilled workers making products by hand and user operated machinery. One-off productions may be commissioned for a 'bespoke' one-off design. These products are often expensive.

2. Batch production

A small-medium quantity of identical products are made at one time. This process can be labour-intensive and will still require skilled workers. However, jigs, templates and moulds are often used to make the process easier and to ensure that all products are exactly the same. It is called batch production because batches of the product can be made as required, with machinery changed easily to suit the required production.

3. Mass production

As its name suggests, this involves making large amounts of identical products. The processes used in this production method are highly automated with few skilled workers. The only workers really required are those who oversee the automated production process to ensure everything is running smoothly.

4. Continuous-flow production

This is similar to mass manufacture, but the number of products created is much larger. The key difference here is that machines run '24/7' (24 hours a day, 7 days a week) to maximise production and minimise costs. Often, no skilled workers are required. With both mass and continuous flow production, the price of machinery, tooling, moulds and maintenance is very expensive, meaning they are only viable for large production runs.

DON'T FORGET

Clean manufacturing is the process of limiting waste, reducing harmful manufacturing by-products and using energy-efficient processes.

VIDEO LINK

Watch the Apple manufacturing process at www.brightredbooks.net

Computer-aided manufacture (CAM)

COMPUTER-AIDED MANUFACTURE

Mass manufacture is run using automated production lines and CNC (computer numerical control) machines/robots. These automated production lines are also known as CAM (computer-aided manufacturing) systems. The **advantages** to the manufacturer of using CAM systems are that machines can:

- work '24/7' non-stop without requiring breaks or becoming ill
- limit mistakes through precise CNC manufacture **(quality assurance)**
- eliminate human error, as machines don't tire or become complacent
- create extremely complex and precise products
- reduce waste through precise production and efficient use of exact quantities of materials **(clean manufacturing)**
- reduce production costs, as they require no heating, lighting or wages, and large volumes of the product can be made continuously and quickly.

CAM systems ultimately speed up production in comparison to using a human workforce. However, they come with some **disadvantages**, as:

- CAM machinery is extremely expensive to purchase
- set-up (installation) and maintenance costs are also very expensive
- malfunction can cause production to stop, costing companies time and money
- a reduction in the human workforce leads to the loss of skilled workers.

This said, however, the high volume of products made means that the **price per unit** can be lowered when sold to the consumer. This is because large volumes of a product can be made on one production run with minimal mistakes. This means that companies can recover the costs of set-up and moulds/dies quickly when selling large quantities to the consumer at a low price while still retaining a profit.

3D PRINTING AND LASER CUTTING

3D printing is an additive process that creates a physical object from a digital design. There are different 3D printing technologies and materials you can print with, but all are based on the same principle: a digital model is turned into a solid 3D physical object by adding material layer by layer. 3D printers vary in quality but are capable of printing complex and high-quality designs quickly, mainly in plastic and some metals. As this technology continues to develop rapidly however, we now have 3D printers that can print food and cell tissues for organs (bio-printing).

Laser cutting is a technology that uses a laser to cut various materials such as wood, plastic and metal. Laser cutting works by directing the output of a high-power laser operated with computer numerical control (CNC), to cut out 2D profiles designed on CAD software from a range of materials. Laser cutting is highly accurate and leaves an excellent finish on the cut edge.

As with all CAM processes each of the above comes with **advantages** and **disadvantages**:

Advantages		Disadvantages	
3D printing	**Laser Cutting**	**3D printing**	**Laser Cutting**
Due to the accuracy of both processes, there is less wastage resulting in cheaper manufacture.		The bigger and better the machine, the more expensive the cost to buy, setup and install.	
Working mechanical parts can be created quickly allowing accurate testing to take place. Individual components can be created and assembled to create working prototypes.		Both processes require specialist software therefore workforces and users need training, Training costs money and takes time.	
Edits to the design can be made quickly based on decisions made during testing. CAD drawings/models can be updated instantly and reprinted or laser cut for further testing.		Mainly used for printing plastics. Dual material printing has not yet been developed with only one material printed at a time.	Some metals such as copper and brass are difficult to laser cut and not recommended.
3D printing is quicker than building physical models from other materials and limits the need and expense of specialist model makers.	Laser cutting makes the difficult job of cutting brittle plastics easier and reduces materials stresses created when cutting with manual techniques, leading to less breakages.	As 3D printers are more available than ever, more and more people are printing leading to an increase in discarded prints that are harmful to the environment.	Not all woods can be cut on laser cutters. Often they require specialist woods such as laser plywood that are very expensive.
Clients and designers can visualise designs quicker allowing for meaningful discussions to take place.		Printing is highly accurate and can be done quickly depending on the quality of the printer.	Lasers cut at very high speeds meaning design can be cut and tested quickly.
CAD models can be rapid prototyped into physical 3D models and prototypes for accurate testing.	Highly accurate and virtually any 2D shape can be cut into materials.	People have been able to print harmful items such as weapons on 3D printers.	

A basic 3D printer

Laser cutter

VIDEO LINK

Check out the clip on FDM rapid prototyping at www.brightredbooks.net

ONLINE

Learn more by following the link at www.brightredbooks.net

DON'T FORGET

You do not need to learn each method of rapid prototyping; but be aware of its uses and advantages.

ONLINE TEST

Test yourself on manufacturing in industry at www.brightredbooks.net

THINGS TO DO AND THINK ABOUT

1. A company is considering setting up a CAM system for their production line. Discuss the advantages and disadvantages of doing this.
2. What are the advantages for the client, designer and manufacturer when using rapid prototyping?

THE IMPACT OF MANUFACTURING ON SOCIETY

As manufacturing technologies improve, so do the products we make. However, this continual advancement in technology can impact positively and negatively on society. For this reason, it is extremely important that you are aware of how manufacturing impacts on society.

Automated car production line

REDUCTION IN WORKFORCE AND THE LOSS OF A SKILLED WORKFORCE

As our demand for high-quality, low-cost products increases, designers and manufacturers have to think about how they can achieve this for us. One way is to use CAM systems, as mentioned previously. More and more, robots and automated production lines are becoming the preferred option for manufacturers, as these ultimately reduce manufacturing costs and speed up the production process. Humans are therefore being replaced by robots, leading to job losses, as a few people can operate an entire robotic production line without the need for specialist skills. This also reduces the need for skilled workers, as computer-control specialists are the only skilled workers required to help maintain the robotic machinery when necessary. Overall, this leads to insecure jobs, the loss of specialist and traditional manufacturing skills, high unemployment rates and the economic decline of industry in villages, towns and countries.

COST OF EQUIPMENT

Setting up a CAM production line is not cheap, as was outlined in the previous chapter. Companies planning to switch to this form of industry must ensure they will gain economic return on their investment. Some workers may also need retraining, which costs money; or companies may have to employ specialist staff to maintain the robots. These workers will require higher salaries – and, even if this maintenance work is outsourced, companies will have to pay a high fee for this service. This can ultimately affect the cost of the end product, driving up prices. This may also be the reason for companies to seek cheaper manufacturing in other countries.

Globalisation and social/moral implications in global manufacturing

The **rise in consumerism** due to more affordable products, and an ever-increasing global population that demands more products, has created a worldwide market for designers and their products. This is good for designers and companies; however, there are problems that **globalisation** has created. Companies are forever looking at ways to reduce manufacturing costs to make more affordable products. As manufacturers look for cheaper ways to build products, they often outsource work to factories in other countries where production costs are lower. Although this is good for companies, it is not always good for the workers and our economy.

Some countries have fewer restrictions on health and safety and workers' rights. Workers in these countries are often subject to lower wages, higher pollution and longer working hours; and sometimes the quality of the end product can suffer, as sufficient regulations are not in place. This also creates a problem for home-based industry. For example, Britain has seen much of its manufacturing industry outsourced to companies in Asia. This has led to the economic decline of Britain's industry and has caused widespread job losses as well as the loss of skilled workforces. There is a large moral issue here that companies must consider, as well as consumers, in ensuring that people the world over are treated fairly. Globalisation also creates environmental problems due to countries having differing standards for green/clean manufacturing, the disposal of waste, transporting products and recycling products.

ENVIRONMENTAL IMPACT AND SUSTAINABILITY

As consumers, we are now more aware than ever of the need to slow down climate change and protect our environment. This has been a positive change in what we want from products, as we want not only products that are of good quality but also products that are more environmentally friendly. Designers, manufacturers and companies therefore have to think about how they can achieve this.

Transport:

- Reducing the size of products and their packaging, or flat-packing products, means that more products can be transported, thereby reducing the number of journeys required.

- Using recyclable/reusable packaging creates less waste.

- Using energy-efficient vehicles will reduce harmful emissions.

Energy efficiency

- Products may be made more energy-efficient to consume less electricity or other fuel.

- Products often come with an energy-efficiency rating to make us aware of this.

- Other ways of powering products, such as solar power, may be investigated.

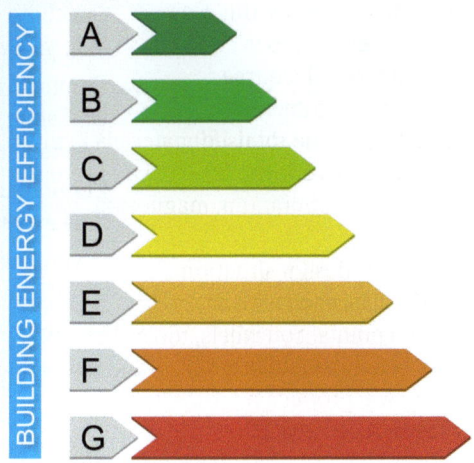

Sustainable materials and manufacturing:

- Using more sustainable or recyclable materials (see also p. 37).

- Using recycled materials or components to build new products rather than raw materials.

- Reducing the amount of components used in products.

- Reducing emissions from manufacturing processes and factories.

- Reducing toxic waste from the by-products of manufacturing processes.

- Recycling waste, such as waste water, to be reused in production.

- Using renewable energies, such as wind and solar energy, to power factories.

- Better planning for manufacturing, so that exact numbers of products are made on time without any delays, waste or damage.

- Ensuring the product can be easily dismantled at the end of its useful life for easier recycling and reuse.

VIDEO LINK

Check out the clips at www.brightredbooks.net for more about recycling materials and improving manufacturing.

ONLINE

Head to www.brightredbooks.net to explore this topic further.

DON'T FORGET

Make sure you can explain globalisation as well as the environmental/sustainability issues surrounding manufacturing.

ONLINE TEST

Test yourself on manufacturing's impact on society at www.brightredbooks.net

THINGS TO DO AND THINK ABOUT

1. Watch the video by IKEA (available on the Digital Zone and using URL http://bit.ly/1zVFEYr). It outlines their approach to improving the issues outlined above. This will provide you with an excellent insight into how companies can improve design and manufacturing to benefit society.

2. Select any product found at home. Think about the issues raised in this chapter, and write down how you could redesign to reduce the product's negative impacts.

EXAM ADVICE

THE ASSIGNMENT

Over the course of National 5 Design and Manufacture, you will learn about and practise skills focusing on the topics covered in this book. This knowledge and these skills will be extremely important when undertaking your course assignment and written exam, where you are expected to demonstrate your understanding of and ability in all areas of the course.

ASSIGNMENT BREAKDOWN

You will undertake two linked assignments set by the SQA. The first assignment focuses on design and is worth 55 of the total 180 marks for the overall course award (30% of your grade). The second assignment focuses on manufacture and is worth 45 of the total 180 marks for the overall course award (25% of your grade).

Design assignment

You will be given a design situation. This design situation will come with three related design briefs, each with a specific focus and given specification points. You will then select the one which interests you most and attempt to solve the problem. To do this, you will produce a design folio with **no more than seven A3 pages**. In undertaking the folio, you are expected to:

- Analyse the design brief and use your understanding of design factors to inform your designs.
- Produce initial ideas using sketching, illustration and modelling techniques.

- Develop your ideas exploring them further and justify your decisions for design changes. You should use sketching, illustration and modelling techniques to help you visualise and explain this.
- Demonstrate understanding of materials and workshop manufacture as it informs your design development and plan for manufacture.
- Continuously evaluate your work as well as produce a final evaluation that evaluates your solution/prototype.

Once you have designed the item you will then have to manufacture for Assignment 2. To do this you are expected to demonstrate the following skills:

- Accurately mark out and manufacture your prototype.
- Be able to select the correct tools for manufacturing tasks and use them correctly and safely.
- Assemble your item correctly.
- Ensure all surfaces are free from marks and an appropriate finish is selected and applied.
- Evaluate the success of your final design.

ASSIGNMENT ADVICE

Design folio

NOTE: Graphics and modelling techniques are crucial throughout the folio.

The marking breakdown for the folio is as follows:

- **Analysing a brief:** 8 marks
- **Generating ideas:** 9 marks
- **Developing ideas:** 9 marks
- **Using models:** 6 marks
- **Using graphics:** 6 marks
- **Planning for manufacture:** 6 marks

Try to decide what you will include across your 7 pages to achieve these marks. You may want to consider the following approach:

PAGE 1: Analyse and research the brief

- Analyse the brief and produce your results on the given SQA proforma sheet. Conducting research will help you plan for ideas, and work out how you will solve the brief and specification given.

PAGE 2–3: Initial ideas

- Come up with a range of diverse ideas using your analysis of the brief and research.
- Make sure your ideas are relevant to the specification.
- Include annotations that discuss the design issues related to design factors.
- You may wish to make reference to possible materials and manufacturing issues.

PAGES 4–5: Development of ideas

- Evaluate your ideas against the spec. You could use an evaluation matrix such as the one shown on p. 18.
- This will allow you to identify which ideas to take forward for development.
- Begin to develop your ideas, **exploring** and **refining** their aesthetics, function, ergonomics and so on, making sure that your ideas evolve into something that is better than the initial design and more fitting to the specification.
- Include annotation that highlights key decisions, and make sure you justify (explain) the reasons why you have made changes/developed your idea.

contd

PAGES 5–6: Technical development

- Select a key development(s) to take forward and begin to analyse exactly what you are going to make, whilst also planning sizes.
- Assess the best materials and manufacturing processes for your chosen development, and compare their advantages and disadvantages.
- You should consider the resources you have available in your school.
- Work out sizes for your design based on things such as items it will hold/store or any restrictions given in the specification. You could also measure the space where it will be used/displayed.

PAGE 7: Planning for manufacture

Using the given SQA proforma sheet you will be expected to:

- include working drawings with dimensions and a cutting list that details all critical sizes for the manufacture and assembly of your design
- produce a sequence of operations to help you plan how you will the stages of manufacture and resources required.

REMEMBER: this is only an example of how you could approach your design assignment and there are many ways in which you could do this.

Manufactured prototype

The marks for manufacture are broken down as follows, and you should also consider the following tips:

Measuring and marking out: 9 marks

- Accurately measure all sizes based on your working drawings.
- Draw lines lightly to save time during finishing work later on.
- Ensure all lines are straight and the correct tools have been used.

Using hand and machine tools: 18 marks

- Make sure cuts, shaping processes or anything else are made accurately to the lines and sizes you have marked out.
- Make sure that joints/fittings/components work as planned, and you have not had to overly rework parts or deviate from your original plan to make them fit.

Assembly of components: 5 marks

- Dry-clamp your work first to make sure everything fits and the model is square, level and true.
- Once it is clamped, double-check your gluing checks (see pp. 60–61).
- Make sure you remove any excess glue using a damp cloth.
- The assembly should be secure, square, level and true in all cases.

Finishing: 9 marks

- Prepare all surfaces of your materials just before assembly. See pp. 46–49 for further help.
- Make sure there are no marks from working or pencil lines.
- For final finishing, ensure you apply your chosen finish correctly and with care, leaving no runs or uneven layers. See pp. 46–49 for further help.
- If using brushes, ensure that no loose bristles become stuck to the surface.

Evaluating: 4 marks

- This can be carried out throughout the manufacture of your design. You should be able to demonstrate the ability to conclude your work and summarise.
- ake sure you evaluate how well you have achieved your planning for manufacture as set out in your design folio. You should also consider how well you have achieved each of the manufacture areas listed above.

DON'T FORGET

The assignments are worth 55% in total, therefore it is extremely important you make every effort to produce work that achieves everything outlined on this page.

DON'T FORGET

The level of independence with which you carry out your work will also affect your mark, therefore you should try working as independently as possible.

ONLINE

More information on the course can be found at www. brightredbooks.net

THINGS TO DO AND THINK ABOUT

To give yourself the best chance of getting full marks, you must plan ahead. Read the brief thoroughly, and make sure you understand everything in the specification. Remember that the approaches outlined above are for advice only. You do not need to follow these, and can take your own approach. Make sure you have consulted your teacher and listened to any instructions you have been given.

QUESTION PAPER

The question paper is set by the SQA and is worth 80 out of the total 180 marks for the overall course award, which is equivalent to 45% of your grade. The exam is broken down into two sections and will seek to test you on your knowledge and understanding of all of the topics covered in this book. The sections are usually broken down as follows:

SECTION 1

Section 1 is worth 60 marks and assesses design and workshop-based manufacture and consists of six or seven questions. Question 1 has **30 marks**. It assesses a range of materials, hand tools and machinery and is based on a workshop-crafted product. This question follows a similar format each year and requires reasoned responses to practical manufacturing tasks. You will be given a product as the sole focus for questioning and asked to answer questions relating to materials choices and the manufacturing techniques used in its production.

For example, you may be shown a manufactured item such as the storage unit shown here.

You may then be given a series of questions that ask you to:

- State what materials are suitable for the wood or plastic components.
- Explain the environmental or manufacturing considerations with regard to using these materials.
- Describe the processes used for manufacturing the item. For example, you would have to describe the process of line bending for manufacturing the plastic component.
- State suitable joining techniques for the wood/plastic, and then describe how these would be carried out. For example, if the question focused on joining the wood and plastic, you may be asked about drilling the plastic so that screws can be used to fix the materials together. You could also be asked to name a suitable wood joint for joining the wooden parts, and describe how to manufacture it. See pp. 56–57 for an example of this type of question.

Make sure you read each question thoroughly so that you understand exactly what is being asked. Remember to answer questions correctly if you are asked to:

- **State:** Use correct terminology to name or state what is being asked for.
- **Describe:** Give a descriptive account or details of processes, steps, properties and so on.
- **Explain:** Describe your answer, but this time with justification and reasons. Think about: how and why?

The remaining questions are worth **30 marks** and assess design as specified in the 'Skills, knowledge and understanding for the course' table. The context of the questions is design work and products that focus on particular aspects of design.

EXAMPLE:

Q: Designers use a range of graphic techniques to communicate their ideas.

State the name of a suitable graphics technique for the stages listed here, and give one reason as to why it would be suitable.

a) Initial ideas (2)

2-point perspective sketching. This technique allows you to visualise the ideas in 3D and gives you a realistic version of how a product would look in real life.

contd

b) Planning for manufacture (2)

Orthographic working drawing. This would allow the manufacturer to see how the different components of the product go together; and dimensions can also be shown.

c) Designers often use models to help them visualise their ideas. Explain the difference between mock-up models and scaled models. (2)

Mock-up models are quick, low-cost and inaccurate. They can be used for assessing the aesthetic qualities of a design. Scaled models are accurate to size, and they function similarly to the real product. Due to scaling, testing of ergonomics can take place and sizes can be worked out.

SECTION 2

Section 2 has 20 marks. This section assesses commercial manufacture and consists of four or five questions. The first question in this section assesses materials and commercial manufacturing processes. This question follows a similar format each year. Candidates identify, select and justify suitable materials and processes for the commercial manufacture of existing products. The remaining questions assess the impact of commercial manufacture on society and the environment and other aspects of commercial manufacture, as specified in the 'Skills, knowledge and understanding for the course' table.

Q: A chair design is shown here. The plastic seat was made using injection moulding.

a) State the name of a suitable material for the seat, and give reasons for your choice. (3)

ABS would be suitable, as it is a thermoplastic suitable for injection moulding. ABS is also hard-wearing and scratch-resistant. ABS comes in a variety of colours, making it suitable for mass manufacture.

b) State two initial set-up costs of injection moulding. (2)

- High tooling costs
- Cost of manufacturing moulds
- Initial set-up cost of machinery.

c) State two features that would help you to identify that the chair had been injection-moulded. (2)

- Injection marks
- Sprue marks
- Ejector pin marks
- Tapered edges/draft angles
- Complex shapes
- Mould split lines
- Single part.

DON'T FORGET

Past papers will help you in preparing yourself for answering exam-style questions.

ONLINE

Past papers can be found by following the link at www.brightredbooks.net

THINGS TO DO AND THINK ABOUT

Try answering the questions you know first of all, and don't panic about the ones you are unsure of. Answering the questions you know will help build your confidence, and you will likely be able to answer these relatively quickly. You can then go back and answer the other questions you were unsure of, giving yourself time to think. Panicking will only make the exam more difficult. Be well prepared, and you will be fine.

APPENDICES

ANSWERS

DESIGN

Problem identification, design briefs, research and specifications 1 (p. 11)

1. Three ways in which opportunities for new product development can be found include:

   1. **Product evaluations:** This involves looking at current products on the market and looking for opportunities to improve them or to produce something much better.

   2. **Market research:** Finding out what consumers need or want helps to identify gaps in the market or potential areas for product development, allowing us to design new products that satisfy consumer needs and wants.

   3. **Technology push:** Researching new manufacturing techniques or technologies can allow designers to incorporate technologies from other industries into commercial designs, which creates new products.

2. A design brief is important because it outlines the problem clearly for the design team and provides a brief outline of what must be done. It also outlines any restrictions the designer must adhere to.

   The design brief must be analysed to allow the design team to identify what must be researched if a good design is to be created, as well as examining all possible options available for creating the solution.

Problem identification, design briefs, research and specifications 2 (p. 13)

1. **Product:** Common hand blender

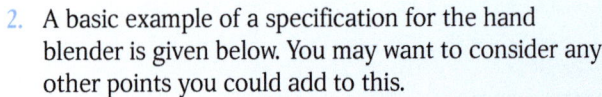

User trial: A user trial could have been carried out, getting people to test the functionality of different hand blenders to see what works best. This data could be used to inform the design of the handle and buttons.

Test rig: Different foods could be blended at different speeds to see if the blades become clogged or if they jam. These tests could all assess what foods the blades struggle to cut at different speeds. This would have helped the designer to ensure that the product performed well in all situations.

2. A basic example of a specification for the hand blender is given below. You may want to consider any other points you could add to this.

   1. Function
   - Must be able to blend different food types easily.
   - Must have different speed settings for blending different food types.

   2. Aesthetics
   - Must look good and fit in with the kitchen environment.
   - Must be made from a durable metal suitable for kitchen use, such as stainless steel.

   3. Ergonomics:
   - The diameter of the body must be designed for the 5th to 50th percentiles of potential users.
   - The blender must be light and portable.
   - The buttons must be easy to press, requiring minimal force.

   4. Safety:
   - The materials must be electrically safe and water-resistant to ensure that liquids and wet foods don't damage the electrical circuits.
   - The materials must be easy to clean for hygiene reasons.
   - The blades should be well secured and protected to reduce the risk of fingers coming into contact with them.

Considering design factors (p. 21)

With the continued rise of consumerism, companies are looking for cheaper ways to design and manufacture their products. This can result in products being made in 'sweat shops' in developing countries with lower-paid workers who do not have the same access to workers' rights. Factories in these countries do not have the same regulations for waste disposal and sustainable production, meaning that harmful by-products can be released into the environment. With the constant development of technology and product updates, consumers are discarding products long before the end of their useful life, meaning that these often end up in landfill, creating more waste and damaging our environment.

Market (p. 27)

Needs can be described as the basic necessities humans require to survive, e.g. food, water and shelter. Wants can be

contd

described as the things that we desire. Although we don't necessarily need these, they can help to satisfy our needs.

Aesthetics (p. 31)

The form of this product is organic as it is based on a tree. However, the lines are quite geometric, providing space for the books. This creates contrast, making the design stand out and look interesting. The shape looks unique due to the design's asymmetric proportions. Finally, the narrow base makes the product look unstable.

Ergonomics (p. 35)

Example answer

Product: Common tin-opener

Anthropometrics: The length of the handle must be longer than the 95th percentile hand width.

This means that all users' hands will fit the handle comfortably. The width of the handles from top to bottom must be within the 5th to 50th percentiles, allowing the majority of users to comfortably hold, grip and squeeze the handles together. The width of the turning lever must be as wide as or wider than the 95th percentile thumb and index-finger size. The length of the turning level should suited to the 50th percentile index-finger length.

Physiology: The turning lever for operating the tin-opener must be easy to turn and require little force. The handle should be easy to squeeze and keep closed when operating the tin-opener. Both of these physiological aspects should be aimed at the 5th percentile's strength, as this will ensure that everyone can easily operate the tin-opener.

Psychology: The red handles and turning lever make it easy for the user to identify the parts of the product they are to interact with and touch.

MATERIALS AND MANUFACTURE

Selecting materials (p. 39)

1. **Standard lengths** are when materials have been pre-cut to off-the-shelf sizes. For example, manufactured boards come in a range of board sizes such as 2400*1200 and 1200*600 mm.

   Standard components are items such as nuts and bolts that are pre-made and can be used in range of different products. These components are interchangeable and save the designers from having to constantly design individual fixings/fittings for products.

2. The **advantages of standard sizes** are that manufacturers can cost manufacturing accurately, which reduces cost and waste. Materials can be purchased straight off the shelf in stores without the need to order specific sizes for every job.

   The **advantages of standard components** are that it makes design and manufacture easier, as not every component has to be specifically designed for a product. It also means parts can be replaced easily if broken, and parts can be made interchangeable for a range of products, for example in flat-pack furniture fittings.

3. **Recycling** involves breaking materials down into a reusable form, such as melting plastic bottles to make new plastic bottles, or plastic materials for clothing such as polyester. Although this is good, recycling produces chemical by-products that harm the environment.

   Upcycling involves reusing working materials/parts from broken products and incorporating them into new products without the need for destructive recycling. This could also include upcycling furniture, for example where old furniture can be restored to look new again. This stops products from ending up in landfill.

Wood (p. 41)

1. Softwoods and hardwoods are both natural woods that come from trees. Softwoods take less time to grow, are readily available and are more sustainable, as they can be easily replanted. Hardwood trees take hundreds of years to grow, and cutting them down is harmful to the environment, as we cannot regrow them easily due to their location and length of time to grow. Hardwoods tend to be stronger due to their tight grain created by the length of time it takes them to grow. Manufactured boards are made from recycled waste. For example, MDF is made from gluing and compressing waste sawdust into large boards.

2. Manufactured boards are more sustainable, as they are made from recycled waste. They come in board size, allowing for bigger products to be made more easily. They can be made to look like expensive hardwoods by applying thin wooden veneers to their faces. This also reduces the amount of hardwood trees that need to be cut down.

Metal (p. 43)

1. Ferrous metals are metals that contain iron, meaning they will rust. Non-ferrous metals do not contain iron and can be described as pure metals. Alloys are a combination of different metals, used to improve the properties of metal such as their aesthetics and durability.

ANSWERS (CONTD)

Plastic (p. 45)

1. Thermoplastic are plastics that return to their original shape when reheated after moulding. This can be done several times, meaning they are particularly useful in manufacture. Thermosetting plastics do not return to their original shape when reheated and are particularly suited to products that require heat resistance or electrical insulation.

2. The seven symbols for plastics are as shown alongside.

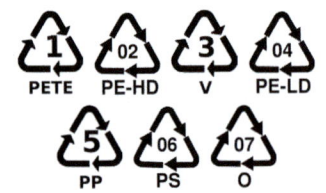

Finishing wood (p. 47)

1. A coarse-grade sandpaper should be used first to remove pencil marks or surface imperfections. Once these are removed, the wood should be further sanded using a smooth-grade paper to smoothen the wood to a good finish ready for the application of a wood finish. It is important when sanding that you sand in the direction of the grain to avoid further damaging the wood and that the sandpaper is wrapped around a cork block to ensure edges remain square.

2. MDF, as it is durable and takes paint easily, which allow the seat to be made any colour. To further improve the durability of the seat and make it more hard-wearing, a wood varnish could be applied on top of the paint.

Wood lathe (p. 59)

1. Begin by marking the diagonals on the ends of the wooden blank to locate the centre. On one end, saw down both lines no more than 5 mm to create slots for the fork to grip. Using a marking gauge, mark parallel lines down the corners of the blank. Use a smoothing plane to remove the sharp corners, which will help to reduce the friction and force required when chiselling on the lathe.

Forming and joining wood 1 (p. 61)

First, you should check the item is **true**, checking that all parts fit as was intended and that edges/surfaces are flush (level) where required.

Check if the item is **square** by using a try square in all corners to check if they are at right angles and by measuring the diagonals.

Ensure the item has not **cupped** because the clamps have been over-tightened. If it has, place weights on top of the product to keep it flat.

Ensure the item hasn't **twisted** by making sure the product is kept flat to the top of the clamps.

Forming and joining wood 2 (p. 63)

1. A knock-down fitting can be described as any standard component used specifically for flat-pack furniture.

2. Flat-pack furniture can be made cheaply for consumers, as there is no need for manufacturers to assemble the item. Standard components (knock-down fittings) also reduce the cost of manufacture, as pre-made components can be bought in bulk. Self-assembly can provide the consumer with a sense of achievement, and anyone can essentially do it. The fact that the item is flat-packed means that storage is easier for the manufacturer; and the consumer can collect the item themselves, taking it home in their car. This also reduces transport costs for the manufacturer, as they can fit more of the product into lorries and so on. This is also beneficial for the environment, as fewer journeys have to be made to transport the item, reducing transport emissions.

Shaping and forming metals (p. 67)

1. **Annealing** is used to heat the metal and allow it to cool naturally. This reduces the tension in the metal, making it easier to work. **Hardening** is where you heat the metal until just before melting point and then quench it in cold water. This causes the molecules in the metal to go into shock, hardening it, which makes it very brittle. **Tempering** is used after hardening. By heating the metal to different temperatures, you can adjust its properties, making it more suitable for specific products.

DON'T FORGET

You could include sketches to help you explain your answer if you find the wording difficult.

contd

Joining metals (p. 75)

1. **MIG welding** melts a reel of filler material that is fed continuously through a welding gun. This is easier **than TIG welding**, where a rod of filler material has to be melted using a gas-flame torch or arc-welding machine. In MIG welding, a continuous bead can be run, making it suitable for larger jobs. Both MIG and TIG welding are good for joining thick metals. **Spot welding** is used for joining thin metal sheets, where a current is passed between two sheets through the copper ends on a spot-welding machine, fusing them together.

2. **Soft soldering** is used for small jobs and requires the use of a soldering iron. It can be carried out manually or using an electric soldering iron. A filler material is melted using the soldering iron to form a join. It is often used for joining thin sheets of metal or creating circuits in electrical engineering. **Hard soldering**, also known as silver soldering, is normally used for plumbing work. Due to the high melting point of the filler material, a gas air torch is used to melt the material and form a join.

Shaping, forming and joining plastics (p. 77)

1. Using a steel rule, engineer's square and permanent marker, measure and mark out the lines where the hook is to be folded. Heat each line to be bent, using a strip heater. Once heated, bend as required and form around a jig if necessary.

2. Heat the plastic in a plastic oven until it is at the right temperature to bend. Once softened, take the plastic out of the oven and twist around a jig designed to the same shape and diameter as the drinking cup. Hold the plastic in place until it sets.

Vacuum forming (p. 81)

1. Moulds for vacuum forming have rounded corners and tapered sides to make it easier to remove the mould from the plastic once formed. Without these, the mould would get stuck inside the plastic due to how tightly vacuum forming forms the plastic around the mould.

2. Due to how tightly the plastic is formed around the mould in vacuum forming, the plastic can stretch over the corners, where the most pressure is applied. As the material stretches over the corners, it can thin. This is known as thinning.

CAM (p. 85)

1. Swapping to a CAM system would mean the manufacturer can leave the machines running 24/7, which would increase production as they don't need sleep like humans. The precise nature of CAM manufacture would limit any mistakes, reducing waste and producing continuously accurate products. This also eliminates the risk of human error. Using CAM systems reduces cost for the manufacturer, as there is no need to heat or light factories, and the precision of machines means exact quantities of materials are used.

   However, the set-up cost of a CAM system is extremely expensive. Any malfunctions in the machinery can stop the entire production line, costing companies time and money. The more manufacturers switch to a CAM system, the more skilled workers become unemployed. This also results in the loss of specialist skills.

2. **Client:** CAD models can be made into physical products, allowing the client to visualise the design better and get hands-on with it. This allows for better evaluation of the product and allows the client to give the designer their insight on what needs to be done.

   Designer: Rapid-prototyped models can be tested due to their accurate representation of the real product. Any changes required can be made easily, and new models can be rapid-prototyped until the design is right.

   Manufacturer: Mechanical fixings and moving parts can be accurately printed using rapid prototyping, meaning that manufacturers can assess the best way to construct and manufacture the product. Rapid-prototyped models can be made using similar materials to the real thing, meaning that accurate testing can take place.

3D printing – a simple and low-cost version of rapid prototyping

accountant – person who oversees project finances and provides advice on spending

advertising – a way of promoting a product

aesthetic – the appearance and styling of products

alloy – a metal made from combinations of different metals

analogy – a way of generating ideas by considering how other products natural or manufactured can inform your design ideas

annealing – a heat treatment for metals which softens the metal, making it more malleable and easier to shape

annotation – labels or short notes that discuss design issues and explain the idea in more detail

anthropometrics – determines the various sizes of all parts of the human body, focusing on differences such as gender, ethnicity and age

branding – (e.g. logos and slogans): tells us who a company is and what they are about

brief – outlines the identified design problem and provides a brief description of what is required to be done

CAM – computer-aided manufacture, i.e. where work is done by computers/automated machines

carcase construction – wooden construction where box forms are created

chamfering – removing a square edge to create an angled edge; can be carried out via various processes used in relation to a given material

clean manufacturing – waste reduction through precise production and efficient use of exact quantities of materials during manufacture

client – person/company who commissions a design team to solve a specific problem

consumer – end user of the product, and person who will buy it

consumer demand – the demand created by consumers for products. This causes market pull, forcing designers to design products that we want. Manufacturers also must ensure they can keep up with the demand for a product.

countersink – a sloped edge drilled (using countersink drill bits) around the top of a pre-drilled hole to allow countersunk screws to sit level with the material's surface

designer – creative member of the design team designing solutions

die-casting – a highly automated and fast metal process used for creating complex components

dimensions – (or sizes): often placed on a working drawing to show the crucial sizes of a product

durability – how well a product can withstand wear, pressure and damage (related directly to materials and strength)

ease of maintenance – how easily a product can be maintained in terms of running costs, repairs and general maintenance such as cleaning

ease of use – how easy a product is to use (related directly to ergonomics)

economist – design-team member who studies market prices and economics, providing crucial information such as current market prices for materials

engineer – uses specialist knowledge of materials and construction to advise the design team on the technical aspects of constructing and producing a product

ergonomics – the science and study of how humans interact with the products they use every day. This helps to determine the best possible solution for a design and how to make products as easy to use as possible.

ergonomist – the person who advises designers to ensure that products fully satisfy end users' ergonomic needs

facing off – lathe process where the end of a piece of metal/plastic is levelled to provide a smooth, flat end surface

fad – a product/style that becomes popular very quickly but soon disappears

fashion – something that is current and popular among consumers, remaining fashionable for a period of time until the next new fashion comes along

ferrous – describing metal that contains iron

fitness for purpose – how well a product performs its intended function

form – the 3D shape of an object

forming – process of creating 3D forms in metal, plastic or wood (e.g. joints in wood, or vacuum forming in plastic)

frame construction – wooden construction where frame forms are created (e.g. window frames)

function – how a product works in terms of its primary function (main job) and secondary functions (additional jobs)

globalisation – the demand and distribution of products across the world, resulting in a worldwide marketplace

hardening – process where heated metal is cooled rapidly in cold water to harden it, making the metal brittle and difficult to work

hardwood – timber that comes from deciduous trees, which take hundreds of years to grow

injection moulding – a highly automated and fast plastic process used for creating complex components

knock-down fittings – pre-made fittings used in flat-pack furniture assembly, making assembly easy for the consumer and reducing costs for the manufacturer

knurling – lathe process where a grooved pattern is created on metal or plastic to provide an area for gripping

laminating – process of gluing strips/planks of softwood/hardwood together to create wider planks

lateral thinking – 'thinking outside of the box', i.e. looking at the problem in a completely different way and trying to design ideas that break the norm

lathe – machine used to turn square forms of material into cylindrical forms (centre lathe for metal/plastic; wood lathe for wood)

lifestyle board – used to collate images that help visualise the target market, looking at consumers' interests, housing, fashion, food choices, likes, dislikes and musical tastes

manufactured boards – wooden boards created using waste materials

manufacturer – provides the designer with advice on the manufacturing equipment and processes required to build the product

market niche – targeting specific groups of people within a market segment, and finding a product to suit their particular needs

market pull – when consumers create large demand for a product to fulfil a need or want. Manufacturers must keep up with this demand to ensure there are enough products, and designers/companies must design products that satisfy this demand.

market researcher – person who conducts market research to provide market data to the design team, ensuring that products being designed will satisfy their intended market

market segment – group of consumers with similar interests and making similar purchasing decisions, giving designers opportunities for product developments within these segments

marketing mix – determining the best way to bring a new product to its intended market by considering the choice of product, price, place of sale and promotional material to ensure the product attracts consumers' attention

mood board – used for collating a series of images that help visualise the theme surrounding a product

morphological analysis – breaking down the problem into a series of variables that will then allow you to try different combinations of each variable to identify potential solutions

obsolescence – where products go out of fashion or become obsolete due to newer technologies (ordinary obsolescence), or where designers plan for products to break down after a period of time so that they can be replaced by new and up-to-date products (planned obsolescence)

orthographic drawing – type of sketch/drawing giving a 2D view of a product, looking straight on at its front, side or top (see also 'third-angle projection')

parallel turning – lathe process where the diameter of a piece of metal/plastic is reduced, working along the length of the material

parting off – lathe process where the finished job is removed from the blank (excess/waste) material

physiology – key area of ergonomics used to determine the human body's physical capabilities (looking at strength, posture, movement, flexibility, reaction speed and muscle control)

plastic dip coating – process of coating metal in plastic to provide a protective finish

plastic memory – thermoplastics' ability to return to their original shape when reheated

prototype – fully working version of the final product

psychology – key area of ergonomics used to determine the psychological aspects of a design so that it is easy for the user to understand how it works

quality assurance – process of quality-checking everything during manufacture to limit any mistakes or time wasted in the production of products

rapid prototyping – CAD/CAM process where designs made using computer-aided design software are then made into 3D models

readily available – how easily accessible a material is and if there is a sufficient stock of it

rendering – adding colour to sketches by applying tone, shade, light and reflection

rivet – any of various types of fixing used to fasten metal/plastics together

rotational moulding – a highly automated, versatile and relatively inexpensive process for producing hollow plastic products of various sizes

sand-casting – traditional and widely used casting process in industry where moulds are made in sand, and molten metal is poured in to form a product

shearing – process of cutting metal to size using a guillotine

situation analysis – method of analysing the internal and external factors surrounding a market or product to help identify new opportunities

softwood – timber that comes from coniferous trees, which grow quickly and can be replanted easily, making them more sustainable

soldering – process of joining metal using solder (filler material); used for light jobs

specification – lists all of the key requirements needed to ensure a product is successful

split mould – type of mould in two halves

standard components – common parts used in products such as screws, brackets, wheels, handles, circuits etc. that can be bought in bulk, reducing cost and saving manufacturers from having to make them themselves

standard lengths – materials or parts that are easily available and cut to a specific size, ready for off-the-shelf purchase and saving manufacturers time and money from having to cut and size the material themselves

step turning – lathe process where different diameters are turned down the length of a piece of metal/plastic, creating a stepped effect

style – a distinctive design that lasts long after its peak period of popularity

taper turning – lathe process where a sloped surface is turned down the length of a piece of metal at a desired angle

tapered edges – sloped edges around the outside of a mould to make it easier to remove

target market – the group of people a product is aimed at

technology push – design strategy where existing products are continually updated and improved to create demand for the latest version. New products can also be developed through scientific research, advances in technology and new materials and advances in manufacturing processes.

tempering – heating hardened metal to different temperatures to alter its properties, making it more suitable for specific jobs

thermoplastics – plastics that return to their original shape when reheated

thermosetting plastics – plastics that do not return to their original shape when reheated but have excellent chemical and heat-resistance properties

thinning – corners or edges of a plastic sheet that thin due to the plastic being stretched tightly over the mould during vacuum forming

third-angle projection – an orthographic drawing showing the elevation, drawn with the end elevations positioned directly to the side of this and the plan view directly above it

thought shower – a group of people quickly bringing together several ideas for solutions good and bad which are then evaluated, and the best idea is taken forward

vacuum forming – plastic process for forming thin sheet plastics into products such as sandwich packaging or chocolate-box trays

welding – process of joining metal and/or plastics together using a filler material (for larger jobs where more strength is required)

working drawing – an engineering drawing that allows the manufacturer to build the product, working with specified dimensions